Applied Regression Analysis for Business

Jacek Welc • Pedro J. Rodriguez Esquerdo

Applied Regression Analysis for Business

Tools, Traps and Applications

 Springer

Jacek Welc
Wrocław University of Economics
Jelenia Gora, Poland

Pedro J. Rodriguez Esquerdo
University of Puerto Rico
Rio Piedras, Puerto Rico, USA

ISBN 978-3-319-71155-3   ISBN 978-3-319-71156-0   (eBook)
https://doi.org/10.1007/978-3-319-71156-0

Library of Congress Control Number: 2017959578

Printed on acid-free paper

This Springer imprint is published by Springer Nature
The registered company is Springer International Publishing AG
The registered company address is: Gewerbestrasse 11, 6330 Cham, Switzerland

Preface

Regression analysis, as a part of statistics and econometrics, constitutes one of the most commonly applied quantitative analytical tool in economics and business. It serves various purposes, from simple inferring about directions and strengths of relationships between economic variables to simulating and forecasting the future. However, it is often abused, either by being applied in areas which by their nature are "immune" to statistical modeling (where obtaining reliable statistical findings or forecasts is often impossible, even though elegant regression models may be built) or by being applied without adequate care devoted to possible traps (e.g., structural changes occurring within modeled relationships or a presence of multivariate outliers).

In this book, we focus on single-equation linear regression models, which seem to constitute the most commonly applied tool of econometric analysis. Our main purpose was to offer a book which is strongly biased toward practical business applications and which is understandable for all those interested in applied statistics and econometrics who are not mathematically inclined. Accordingly, we do not deliver any mathematical proofs. Also, we provide only those formulas which are necessary to understand and apply the individual tools discussed in our book. We believe therefore that our book will be "digestible" for a broad universe of non-statisticians who are interested in applying regression analysis. However, the basic knowledge of the nature, structure, and purposes of the single-equation regression models is a prerequisite to understand the issues covered by this book.

Our business practice has taught us that unskilled and mechanical application of regression analysis may be very hazardous and may result in poor business decisions. Therefore, a significant part of this book is devoted to discussing and illustrating the most typical problems and pitfalls associated with applying regression analysis in a business practice. Two chapters are fully dedicated to such issues: Chapter 2, which discusses the problems arising from outlying and influential observations, and Chap. 6, which deals with several other (often somewhat subtle) traps. To our knowledge, some of the issues covered by those two chapters are overlooked by the majority of other books on applied statistics and econometrics.

To be reliable, any regression model must satisfy several conditions. Accordingly, an important step in building such models is a verification of their correctness and statistical properties. These issues are discussed with details in Chap. 4, which not by chance is the longest one in our book. To make this subject more understandable and convincing, apart from presenting the individual tools for verifying the model we also deliver simple illustrative examples of causes and effects of the most common flaws of the model (such as autocorrelation, heteroscedasticity, and lack of normality of regression residuals). We also discuss two issues which are not covered by the majority of other books, i.e., randomness and symmetry of regression residuals (with simple statistical tests).

Our book is organized as follows. Chapter 1 offers a brief reminder of basics of regression models. Chapter 2 illustrates common problems resulting from the presence of outlying and influential observations in a sample and offers several tools useful in detecting them. In the following two chapters, we discuss the individual phases of the regression model building, including selection of explanatory variables and model verification. Chapters 5 and 6 discuss the most common simple adjustments to regression models (such as dummy variables) and illustrate problems and traps caused by unskilled estimation and application. In Chap. 7, we briefly present the two classes of models useful in analyzing discrete dependent variables, that is, discriminant analysis and logit function. The book closes with two chapters which offer real-life examples of the single-equation linear model building and application, of which one deals with time series while another one with cross-sectional data.

Jelenia Gora, Poland   Jacek Welc
Rio Piedras, Puerto Rico   Pedro J. Rodriguez Esquerdo

Contents

Chapter 1
Basics of Regression Models

1.1 Types and Applications of Regression Models

A regression model, also called an econometric model, is a quantitative analytical tool in which the behavior of some variables is explained by other variables. A single-equation regression model has the form of an equation (a mathematical function) which quantifies a relationship between a dependent variable (which is explained by the model) and one or more explanatory variables, which are also called regressors (and which have statistical or causal relationships with the dependent variable).

In a theoretical notation, which does not assume any specific functional (mathematical) form, the econometric model may be specified as follows:

$$Y = f(X_1, X_2, \ldots X_i, \varepsilon),$$

where:

Y—dependent variable,
$X_1, X_2, \ldots, X_i$—set of explanatory variables,
ε—residuals from the model (i.e., that part of a variability of the dependent variable which is not explained by the model).

The simplest and probably the most commonly applied functional form of the regression model is a linear one, which has the following general form:

$$Y = \alpha_0 + \alpha_1 X_1 + \alpha_2 X_2 + \ldots + \alpha_i X_i + \varepsilon,$$

where:

Y—dependent variable,
$X_1, X_2, \ldots, X_i$—set of explanatory variables,
ε—residuals from the model,

© Springer International Publishing AG 2018
J. Welc, P.J.R. Esquerdo, *Applied Regression Analysis for Business*,
https://doi.org/10.1007/978-3-319-71156-0_1

$\alpha_0, \alpha_1, \alpha_2, \ldots, \alpha_i$—coefficients (structural parameters) of the model which quantify a direction and strength of the statistical relationships between individual explanatory variables and the dependent variable.

Depending on the various criteria, regression models may be classified as follows:

1. According to the number of equations:

   - Single-equation models, where only one dependent variable is explained by a set of explanatory variables in a single equation.
   - Multiple-equation models, where two or more dependent variables are explained in two or more equations.

2. According to the number of explanatory variables:

   - Univariate models, where a behavior of the dependent variable is explained by only one explanatory variable.
   - Multivariate (or multiple) models, where two or more explanatory variables occur.

3. According to the functional (analytical) form of the model:

   - Linear models, where linear relationship between dependent variable and explanatory variables is applied.
   - Nonlinear models, where mathematical functions other than linear are applied (e.g., power or exponential functions).

4. According to the nature of relationships between the dependent variable and the explanatory variables:

   - Causal models, where only those explanatory variables are used which have well-grounded causal relationships with the dependent variable.
   - Symptomatic models, where the set of regressors includes the variables which show statistically significant correlations with the dependent variable, but not necessarily causal relationships.

5. According to the nature of the dependent variable:

   - Models for continuous variables, where the dependent variable is measured on a continuous scale (e.g., inflation expressed in percentages).
   - Models for discrete variables, where the dependent variable usually has a qualitative nature and is expressed in a binary way.

6. According to the measurement basis of the variables included in the model:

   - Time-series models, where individual observations of variables change in time (e.g., a model of a quarterly profit of a company, where individual observations differ from quarter to quarter).

- Cross-sectional models, where individual observations of variables are measured at the same moment of time, but for different objects (e.g., a model capturing the statistical relationship between crime rates in 30 European countries and GDP per capita of those countries, where individual observations of variables are measured for the same point in time, but vary between individual countries).

The distinction between time-series and cross-sectional data is very important for regression analysis, because it affects both the methodology applied in model estimation as well as the practical application of the estimated models. Also, the practical problems (e.g., related to outlying observations) differ between these two types of econometric models.

The regression models are usually applied for the following purposes:

- Inferring about existence (or lack) of relationships between variables (e.g., a researcher may be interested in confirming an existence of the supposed relationship between money supply and inflation rate).
- Forecasting (e.g., an analyst may predict the operating profit of a given company in the next quarter, on the basis of the regression model which explains the behavior of that company's quarterly profit in the prior 20 quarters).
- Scenario analyses (e.g., a central banker may simulate the required change of the market interest rate which is necessary to keep the inflation in the country under control, on the basis of the regression model which captures the relationship between the inflation as the dependent variable and interest rate as the explanatory variable).

The procedure of regression model building, which we will follow throughout this book, consists of the following consecutive phases:

1. Specification of the dependent variable, whose behavior is to be explained by the model.
2. Specification of the set of candidate explanatory variables (i.e., the set of variables which are supposed to have statistical relationships with the dependent variable).
3. Preliminary analysis of the collected data and identification of outlying observations.
4. Adjustment of the set of data for the identified outlying observations (e.g., by removing them).
5. Elimination of the redundant (insignificant) explanatory variables from the set of candidate explanatory variables, by means of:

   - Statistical methods (e.g., the procedure of "general to specific modeling").
   - Economic theory, expert knowledge, or subjective judgment.

6. Final estimation of model parameters.
7. Verification of the model correctness and model fit to the actual data.
8. Application of the obtained model.

1.2   Basic Elements of a Single-Equation Linear Regression Model

In this book, we focus on single-equation linear regression models, which is probably the most commonly applied class of econometric models. The primary elements of such models are:

1. Coefficients (structural parameters) of the model:
   - Intercept, which in the case of models with single explanatory variable is interpreted as the intersection of a regression line with a vertical axis of a scatterplot (i.e., the expected value of the dependent variable at zero value of the explanatory variable).
   - Slope coefficients, associated with individual explanatory variables and which quantify direction and size of impact of a given explanatory variable on the dependent variable.
2. Standard errors of individual structural parameters, which determine the expected ranges into which the true values of individual coefficients (which may not be exactly equal to the estimated ones) fall.
3. T-statistics of individual coefficients, computed by dividing the value of a given coefficient by its respective standard error, which are used in evaluating the statistical significance of individual coefficients (and thus also a statistical significance of individual explanatory variables).
4. Coefficient of determination, also called R-squared, which measures goodness of fit of the estimated model to the actual data.

One of the fundamental principles of the econometric model building is to include in the model only those explanatory variables which show statistically significant relationships with the dependent variable. Thus, estimating a regression model requires evaluation of the statistical significance of its individual explanatory variables. The most commonly applied tool for checking the statistical significance of individual explanatory variables is a test best on t-Statistics computed for those variables (Cox and Hinkley 1974). The procedure of this test is described in Table 1.1.

The coefficient of determination is computed with the following formula (Patterson 2000):

$$R^2 = 1 - \frac{\sum_{t=1}^{n} e_t^2}{\sum_{t=1}^{n} (y_t - \bar{y})^2},$$

where:

R^2—coefficient of determination (R-squared) of a given model,
n—number of observations used in estimating a given model,
y_t—actual observations of the dependent variable,

Table 1.1 The procedure of Student t test for the statistical significance of the individual coefficients of the regression model

Step 1	Formulate the hypotheses: $H_0 : b_i = 0$ (tested coefficient of the model does not statistically differ from zero) $H_1 : b_i \neq 0$ (tested coefficient of the model is different than zero)
Step 2	Compute the empirical statistic t_e: $t_e = \frac{b_i}{S(b_i)}$, where: b_i—tested coefficient $S(b_i)$—standard error of the tested coefficient
Step 3	Obtain the critical value of t_t from the t-Student distribution table for $n - k$ degrees of freedom and for the assumed level of significance α, where: n—number of observations used in estimating a model k—number of explanatory variables in a model
Step 4	Compare the absolute value of the empirical statistic t_e with critical t_t If $|t_e| \leq t_t$, then H_0 is not rejected, which means that the tested coefficient is not statistically different from zero If $|t_e| \geq t_t$, then H_0 is rejected (and H_1 is true), which means that the tested coefficient is not statistically different from zero

Source: Authors

$\hat{y}_t$—fitted (i.e., obtained from a given model) observations of the dependent variable,

e_t—residuals of the model $\left(e_t = y_t - \hat{y}_t \right)$,

$\bar{y}$—arithmetic average of all the observations of the dependent variable.

The coefficient of determination is computed for models which include an intercept and is bounded by a range between zero and unity. Its extreme values (extremely rarely met in practice) may be interpreted as follows:

- R-squared equaling zero means that the model does not explain any part of the observed variability of the dependent variable (which is equivalent to concluding that there is no any statistical relationship between the dependent variable and explanatory variables included in the model).
- R-squared equaling unity means that the model perfectly (without any residuals) explains the observed variability of the dependent variable (which is equivalent to concluding that there exists a deterministic relationship between the dependent variable and explanatory variables included in the model).

In applications of regression analysis to social sciences (e.g., economics, business studies, sociology, or psychology), the extreme values of the coefficient of determination (i.e., zero or unity) are next to impossible. However, it is possible to find models in which case the R-squared is very low (slightly above zero, but with statistically significant t-Statistics of explanatory variables) or very high (slightly below unity). Generally speaking, the higher the R-squared, the better (provided, however, that the model is free from problems such as multicollinearity, discussed later in this book). This is so because the higher the value of the R-squared, the higher portion of the observed variability of the dependent variable is captured and explained by the model.

The coefficient of determination is sometimes used for comparing alternative regression models. It is assumed that from two or more competing models for the same dependent variable, a researcher should choose and apply the one with the highest R-squared. However, it is valid only if the alternative models do not differ in terms of the number of explanatory variables and the number of observations (Goldberger 1991). Otherwise, their coefficients of determination are not directly comparable. In cases where the compared models vary in the number of variables or observations, the adjusted R-squared should be applied. It is computed as follows (Charemza and Deadman 1992):

$$\hat{R}^2 = 1 - \frac{n-1}{n-k}\left(1 - R^2\right),$$

where:

$\hat{R}^2$—adjusted coefficient of determination of a given model,
R^2—"traditional" coefficient of determination (R-squared) of a given model,
n—number of observations used in estimating a given model,
k—number of explanatory variables in a model.

The value of the adjusted R-squared cannot exceed unity. However, it does not have a lower bound, which means that its negative values are possible (in contrast to the "traditional" coefficient of determination).

In the following chapters, we discuss the details of individual steps of the single-equation regression model building and application. We begin with a discussion of the problems caused by a presence of outlying observation in samples.

References

Charemza WW, Deadman DF (1992) New directions in econometric practice. General to specific modelling, cointegration and vector autoregression. Edward Elgar, Aldershot
Cox DR, Hinkley DV (1974) Theoretical statistics. Chapman and Hall, London
Goldberger AS (1991) A course in econometrics. Harvard University Press, Boston
Patterson K (2000) An introduction to applied econometrics. A time series approach. Palgrave, New York

Chapter 2
Relevance of Outlying and Influential Observations for Regression Analysis

2.1 Nature and Dangers of Univariate and Multivariate Outlying Observations

As the name suggests, the outlying observations (commonly called "outliers") are the ones which "lie out" from the remaining observations in the sample. In regression models, the outliers are observations with large residuals (Makridakis et al. 1998). First, we will discuss the nature of the outliers occurring in the case of the single variables, which we will call the univariate outliers. Then we will proceed to discussing the more subtle type of outliers, which occur in the case of multivariate analyses.

2.1.1 Univariate Outlying Observations

Imagine the sample of observations of the stature of ten persons, as illustrated in Fig. 2.1. The simple visual inspection of these data immediately suggests that the tenth person in this sample is outstandingly tall (as compared to any other person within the same sample) and can be described as an outlying observation. Usually an outlier occurring within a single variable is quite easy to detect, even by such a simple visual inspection.

The presence of such outliers, if they are clustered on only one side of a variable distribution (here on the upper side), results in overstated arithmetic mean within the sample. In the case of the above data, the arithmetic mean for all ten observations, equals 176,10 cm. However, there are only three persons with the stature above that arithmetic mean, while seven remaining persons are lower than the mean. If the mean is to reflect the approximate average level of a larger group, then the computed mean of 176,10 cm seems to be biased upward. In

© Springer International Publishing AG 2018
J. Welc, P.J.R. Esquerdo, *Applied Regression Analysis for Business*,
https://doi.org/10.1007/978-3-319-71156-0_2

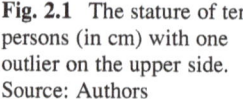

Fig. 2.1 The stature of ten persons (in cm) with one outlier on the upper side. Source: Authors

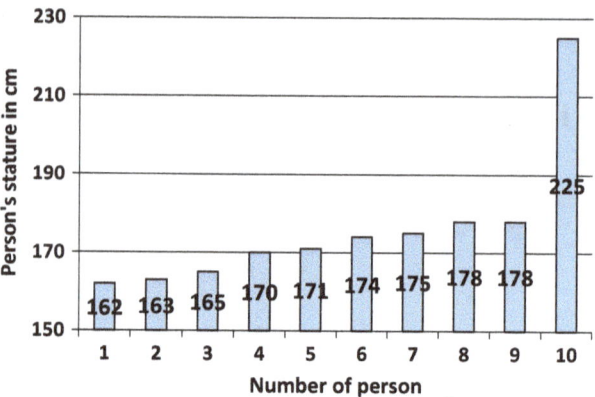

contrast, the arithmetic mean calculated for the first nine persons (excluding the outlying observation) equals 170,67 cm. Because it is very common for economic datasets to include some outlying observations, the economists tend to prefer the median to the arithmetic mean as the measure of a central tendency. In the above sample, the median is the same and equals 172,50 cm, whether computed on the basis of the whole dataset (ten observations) or after excluding the tenth observation.

In the above example, featured by the presence of an outlying observation on one side of the distribution, the arithmetic mean based on a full sample was biased upward and deviated from the sample median. Therefore, if the sample mean deviates significantly from the sample median, it is the first suggestion of the presence of some outliers. If the arithmetic mean exceeds the median by a substantial margin, it is a signal of a probable presence of outliers on the upper side of a sample distribution. In contrast, the arithmetic mean, when it is substantially smaller than the median within the same sample, suggests a probable presence of some outliers on the lower side of the distribution. However, even if the arithmetic mean is exactly equal to the median, it does not mean that the sample is free from outliers. Quite the reverse: it might be that there are even quite extreme outliers on both sides of the sample distribution, but they offset each other. This is illustrated in Fig. 2.2.

As is clearly discernible in Fig. 2.2, now there are at least two outlying observations. However, they are located on the extreme ends of a sample distribution. There is one very tall person and the another one which is uniquely short. In such cases, the outliers on both sides of a distribution tend to partially or fully offset themselves. As a result, an arithmetic mean within the sample might be very close to the sample median. In fact, in the case of the above data, the full-sample arithmetic mean equals 172,50 cm and is identical to the full-sample media. Therefore, comparing sample mean and median should be treated as only first and rough check for the presence of outliers (because it is unreliable if the outliers occur on both sides of a distribution).

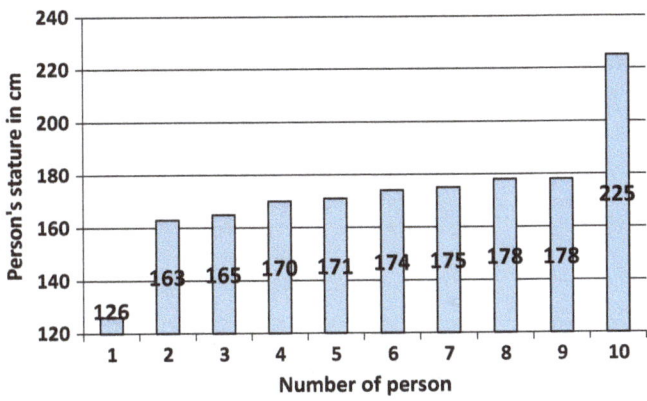

Fig. 2.2 The stature of ten persons (in cm) with one outlier on the upper side and one outlier on the lower side. Source: Authors

2.1.2   Influential Observations Related to Univariate Outliers

Now, we will illustrate the most common distortions of the regression analysis in the presence of the significant outliers. First, we will present the problems caused by the so-called influential or dominating observations. Suppose that for the same ten persons whose stature was shown in Fig. 2.1, we have the accompanying data about their weight. These data are presented in Table 2.1.

As is clearly visible in Table 2.1, the tenth person not only is relatively tall, but is also uniquely heavy (as compared to the other nine members of the sample). Imagine that our task is to explore the statistical relationship between peoples' stature and weight. If we regress the weight of the above ten persons (as the dependent variable) against their stature (as the explanatory variable), we obtain the results presented in Fig. 2.3.

The simple regression presented in Fig. 2.3 is strong, with its R-squared of 0,952. Such a high coefficient of determination means that peoples' stature is an almost monopolistic driver of peoples' weight. However, the visual inspection of the above figure immediately suggests that the obtained results are probably heavily biased by one observation (the person who is both very tall and uniquely heavy). If we remove that outstanding observation and reestimate the regression, we will obtain the results shown in Fig. 2.4.

Now the parameters of the obtained regression look slightly different. The coefficient of determination is no longer that high and now equals only 0,584 (instead of 0,952). It seems more logical because persons' stature constitutes only one (although very relevant) of many drivers of person's weight. But the other regression parameters have changed as well. Particularly, the slope coefficient equals 0,959 and is lower than before by more than 20%.

Table 2.1 The stature (in cm) and the weight (in kg) of ten persons with one outlier on the upper side

Number of persons	Stature (in cm)	Weight (in kg)
1	162	65
2	163	60
3	165	75
4	170	67
5	171	72
6	174	70
7	175	80
8	178	76
9	178	85
10	225	140

Source: Authors

Fig. 2.3 The relationship between persons' weight (in kg) and persons stature (in cm), estimated on the basis of all ten observations presented in Table 2.1. Source: Authors

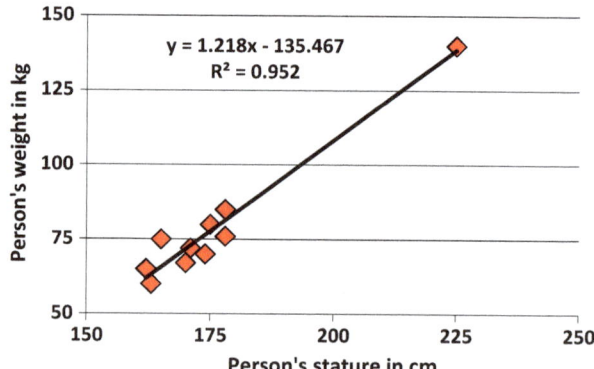

Fig. 2.4 The relationship between persons' weight (in kg) and persons' stature (in cm), estimated on the basis of the observations presented in Table 2.1 after excluding the tenth observation. Source: Authors

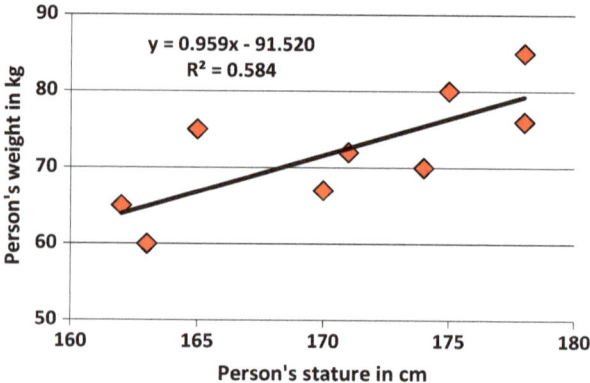

Fig. 2.5 The statistical relationship between a dependent variable (denoted as Y) and an explanatory variable (denoted as X), in the presence of a very strong influential observation. Source: Authors

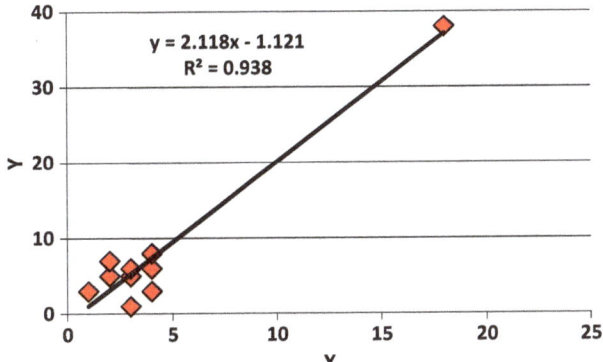

Fig. 2.6 The statistical relationship between a dependent variable (denoted as Y) and an explanatory variable (denoted as X), presented in Fig. 2.5, after excluding an influential observation. Source: Authors

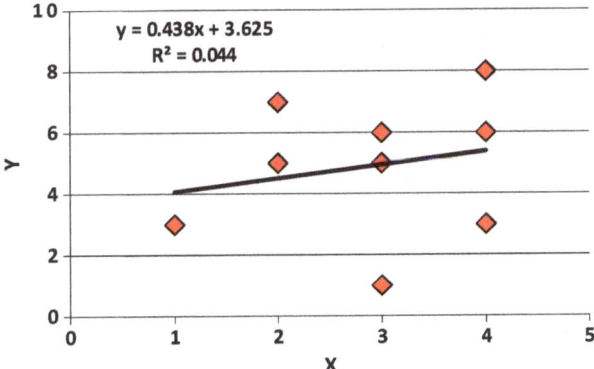

In the above example, the tenth person, who is both very tall and relatively heavy, constitutes the influential observation. In Fig. 2.3, we can see that the tenth person is unique in both dimensions: it has an outlying stature as well as an outlying weight. However, although this outlying observation overwhelms the rest of the sample, it only overstates the R-squared and the slope coefficient. The expected positive relationship between the stature and the weight, shown in Fig. 2.3, holds in Fig. 2.4, although in a weaker form. However, influential observations are sometimes capable of changing a non-existent relationship (for instance, between two completely unrelated variables) into the allegedly strong one, if the influential observation is lying distinctively far from the cluster of remaining observations. This is illustrated in Figs. 2.5 and 2.6.

As both figures show, the allegedly strong relationship shown in Fig. 2.5 (with its R-squared of 0,938) virtually disappeared after removing the influential observation from the sample. Also, the slope coefficient in the regression presented in Fig. 2.6 constitutes only slightly more than 20% of its original value shown in Fig. 2.5. These examples clearly illustrate pitfalls related to outliers and emphasize the relevance of the detection and correct treatment of the outlying observations.

2.1.3 Multivariate Outlying Observations

The univariate outliers, discussed so far, are often easily detectable. In many cases, we can immediately discern some suspiciously high or low values in a dataset of a single variable. The situation is much more complex in case of multivariate outliers. These are the observations which lie distinctively far from the core relationship between two or more variables in spite of having the values of individual variables which are not deemed to be the univariate outliers.

If we look at the data presented in Table 2.2, we may see an another set of data related to peoples' stature and weight. The observations in that table are sorted in order of increasing stature. If we track the first 20 observations, we may notice that they are deterministically (not only stochastically) related. The increase of a stature by 1 cm is associated with the increase of a weight by exactly 1 kg. However, the last observation in the sample is clearly unique: that is someone who is relatively tall (one of the two tallest persons in the sample) while being relatively light (one of the five lightest persons in the sample). If we now look at Figs. 2.7 and 2.8, we might see that none of the observations of the individual two variables (when separated from each other) seem to be an outlier. Further in the chapter we will show that this supposition is confirmed by the commonly applied tools for detection of the univariate outliers (such as three-sigma rule). However, Fig. 2.9 evidently

Table 2.2 The stature (in cm) and the weight (in kg) of a group of persons without any outlying stature and outlying weight

Number of persons	Stature (in cm)	Weight (in kg)
1	161	61
2	162	62
3	163	63
4	164	64
5	165	65
6	166	66
7	167	67
8	168	68
9	169	69
10	170	70
11	171	71
12	172	72
13	173	73
14	174	74
15	175	75
16	176	76
17	177	77
18	178	78
19	179	79
20	180	80
21	180	64

Source: Authors

Fig. 2.7 The stature (in cm) of a group of persons, based on the data from Table 2.2. Source: Authors

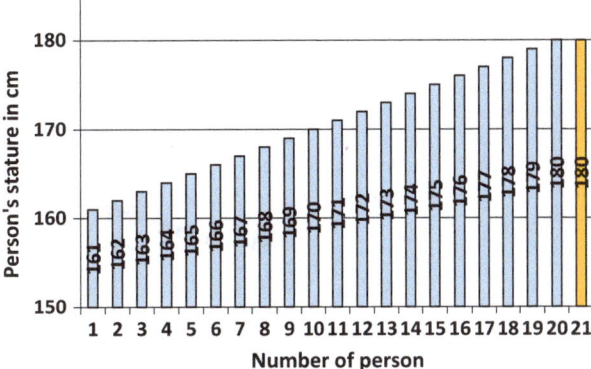

Fig. 2.8 The weight (in kg) of a group of persons, based on the data from Table 2.2. Source: Authors

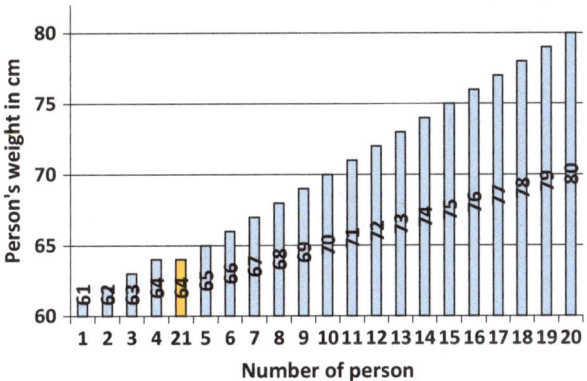

Fig. 2.9 The statistical relationship between the weight (in kg) and the stature (in cm) of a group of persons, based on the full sample of data presented in Table 2.2. Source: Authors

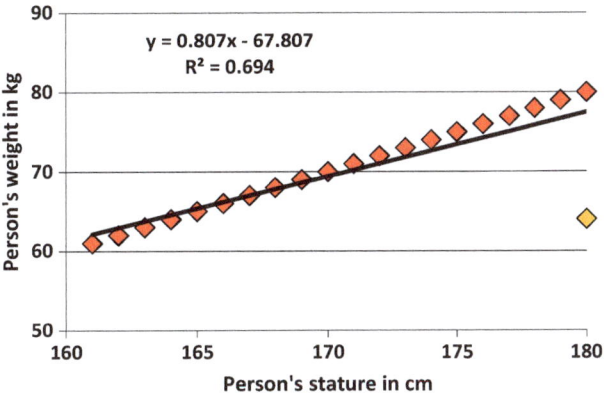

Fig. 2.10 The statistical relationship between the weight (in kg) and the stature (in cm) of a group of persons, based on the sample of data presented in Table 2.2 after excluding the last observation. Source: Authors

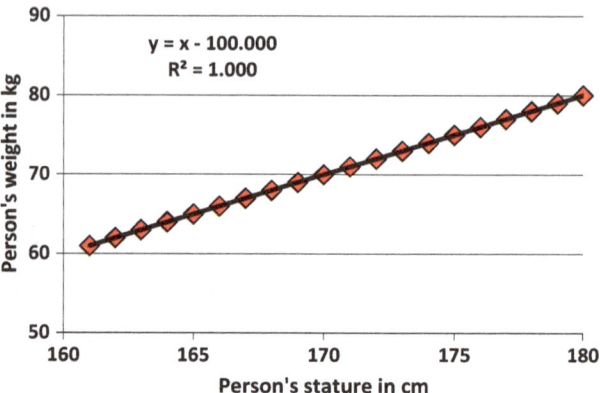

shows that the statistical relationship between both variables may no longer be considered free of outliers. Clearly the last observation from Table 2.2 lies outstandingly far from the core relationship between the two variables.

The outlying observation evidently visible in Fig. 2.9, when included in the sample on the basis of which the regression of the weight against the stature is estimated, heavily distorts the regression parameters. If we compare the regressions shown in Figs. 2.9 and 2.10, we can see that their structural parameters as well as their coefficients of determination differ significantly. Within the subsample of the first 20 observations the relationship is deterministic, thus its R-squared equals unity (instead of only 0,694, as shown in Fig. 2.9). Moreover, in that subsample of 20 observations the increase of the stature by 1 cm is associated with the increase of the weight by exactly 1 kg, thus the slope parameter equals unity (instead of 0,807, as shown in Fig. 2.9).

We remember that the univariate outliers may often be easily detected by means of simple charts which show the values of the individual observations of a given single variable. Such a visual inspection of the single variables did not work in the case of a bivariate relationship, as discussed above for the regression of the weight against the stature, based on all the observations from Table 2.2. However, even in that case we were able to catch the outlier by the visual inspection of a respective scatterplot of the investigated bivariate relationship. Unfortunately, it was possible only because of the bivariate nature of that relationship, whereby we could visualize it on a two-dimensional chart. As will be shown below, the bivariate charts between two variables may be misleading in the case of multivariate relationships, for example, when building the multiple regression models.

Suppose that now we have the data about the weight (denoted as WEIGHT) of a group of persons (which is our dependent variable) and three drivers (explanatory variables) of that weight:

- Stature (in cm), denoted as STATURE.
- Average daily calories consumption (in kcal), denoted as CALORIES.
- Average weekly sporting activity (in hours), denoted as SPORT.

Table 2.3 The weight (in kg) and its three drivers of a group of persons

Number of persons	WEIGHT (actual weight in kg)	STATURE (stature in cm)	CALORIES (average daily calories consumption in kcal)	SPORT (average weekly sporting activity in hours)	FITTED WEIGHT (in kg)[a]	Residual[b]
1	83,1	161	2.500	0	78,73	4,32
2	89,1	162	2.700	0	83,92	5,18
3	95,7	163	3.000	1	89,86	5,79
4	58,2	164	2.000	4	60,76	−2,56
5	62,3	165	2.300	6	65,01	−2,76
6	78,8	166	2.600	3	77,69	1,11
7	72,9	167	2.400	3	73,14	−0,29
8	60,4	168	1.900	2	62,97	−2,57
9	74,5	169	2.200	0	73,96	0,49
10	56,5	170	2.600	12	63,79	−7,29
11	84,6	171	3.200	8	85,47	−0,92
12	80,1	172	2.800	5	81,10	−1,00
13	83,2	173	2.900	5	83,85	−0,70
14	79,2	174	2.600	3	80,24	−1,04
15	56,8	175	2.100	6	63,32	−6,57
16	68,8	176	3.000	12	75,44	−6,64
17	72,9	177	2.800	8	77,63	−4,78
18	80,9	178	2.400	0	81,69	−0,79
19	79,0	179	2.500	2	81,07	−2,12
20	86,0	180	2.900	4	87,76	−1,76
21	87,0	180	2.400	12	62,10	24,90

Source: Authors
[a]From the regression of WEIGHT against its drivers, of the following form: WEIGHT = −33,292 + 0,318 × STATURE + 0,024 × CALORIES − 1,686 × SPORT
[b]Actual weight − fitted weight

These data are presented in Table 2.3. The table also contains the fitted values from the linear regression of a weight against its three drivers as well as the residuals from that regression.

Figure 2.11 plots the residuals from the regression of WEIGHT against its three explanatory variables (on the basis of the data presented in Table 2.3). As can be clearly seen, the last residual in the sample seems to be abnormally large, as compared to the remaining residuals. The value and the direction of that residual means that the actual weight of that person exceeds the predicted weight (from the estimated regression) by as much as almost 25 kg. For comparison, in the case of the remaining 20 persons in the sample, the differences between their actual and fitted weights lie within the range −7,3 kg and 5,8 kg. Thus, the last observation in the sample may be safely called an outlier. Table 2.4 shows why it is important to

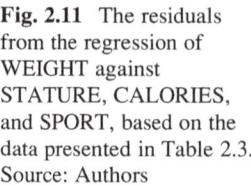

Fig. 2.11 The residuals from the regression of WEIGHT against STATURE, CALORIES, and SPORT, based on the data presented in Table 2.3. Source: Authors

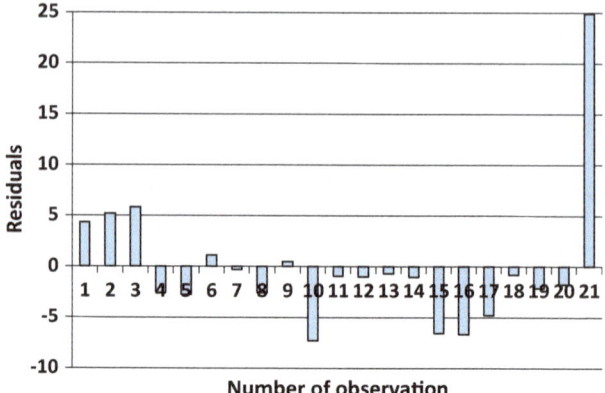

Table 2.4 Slope parameters and coefficients of determination of the two regressions of WEIGHT against STATURE, CALORIES, and SPORT, estimated on the basis of data from Table 2.3

	Regression estimated on all 21 observations from Table 2.3	Regression estimated on the first 20 observations from Table 2.3
Intercept	−33,292	0,000
STATURE	0,318	0,050
CALORIES	0,024	0,030
SPORT	−1,686	−2,500
R-squared	0,654	1,000
Adj. R-squared	0,593	1,000

Source: Authors

be aware of the presence of such an outlier within the sample, on the basis of which the regression is estimated. In this specific example, the relationship between WEIGHT and its three drivers is deterministic within the first 20 observations (it was deliberately designed to have this artificial feature). This can be inferred from the R-squared and the Adjusted R-squared of that regression, both equaling unity. In contrast, adding this outlying observation dramatically distorts the regression results.

Now let's try to identify the potential outliers by looking at the three scatterplots which visualize the relationships between the dependent variable (WEIGHT) on one side and its individual drivers (STATURE, CALORIES, SPORT) on the other side. These relationships are illustrated in Figs. 2.12, 2.13, and 2.14. For comparison, Fig. 2.15 presents the relationship between the actual observations of the dependent variable (data from the second to the left column of Table 2.3) and its fitted values, obtained from the regression of WEIGHT against the three explanatory variables (data from the second to the right column of Table 2.3). None of the observations shown in Figs. 2.12, 2.13, and 2.14 seem to lie far enough from the respective core relationship to suggest the presence of an outlying observation. In contrast, even the casual glance at Fig. 2.15 immediately points to the one outstanding point of a dataset. This is exactly the last observation from Table 2.3.

Fig. 2.12 The statistical relationship between the weight (in kg) and the stature (in cm) of a group of persons, based on the sample of data presented in Table 2.3. Source: Authors

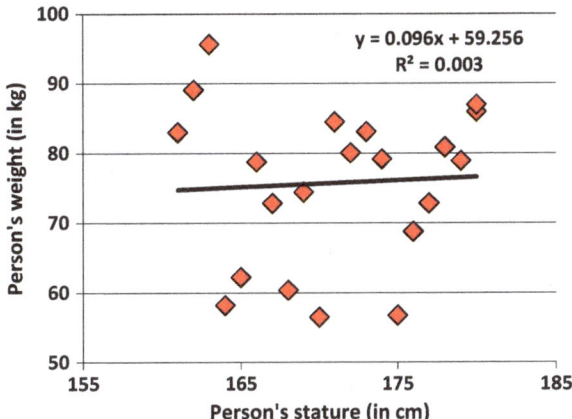

Fig. 2.13 The statistical relationship between the weight (in kg) and the daily calories consumption (in kcal) of a group of persons, based on the sample of data presented in Table 2.3. Source: Authors

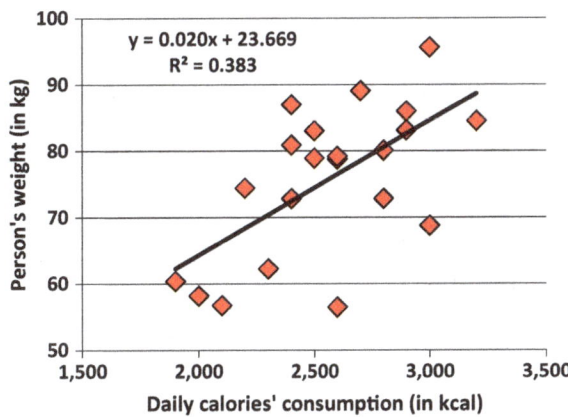

Fig. 2.14 The statistical relationship between the weight (in kg) and the weekly sporting activity (in hours) of a group of persons, based on the sample of data presented in Table 2.3. Source: Authors

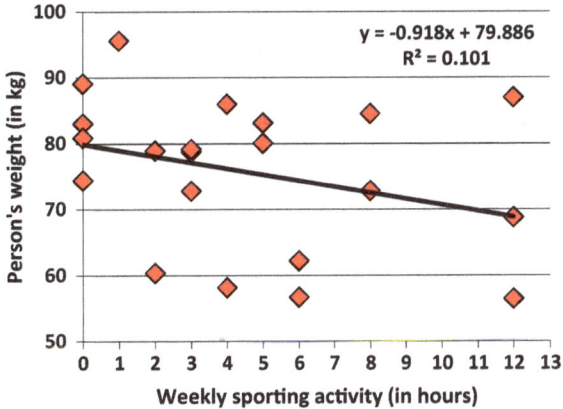

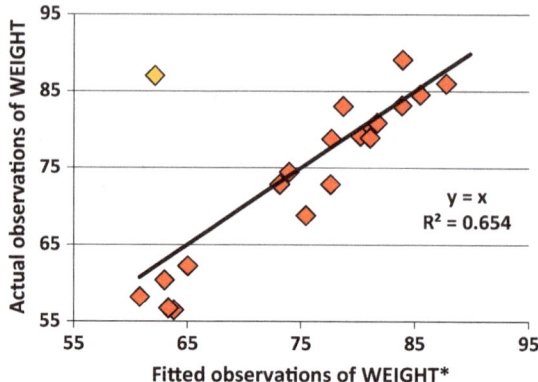

Fig. 2.15 The relationship between the actual weight and the fitted weight, based on the sample of data presented in Table 2.3, from the regression of WEIGHT against its drivers, of the following form: WEIGHT $= -33,292 + 0,318 \times$ STATURE $+ 0,024 \times$ CALORIES $- 1,686 \times$ SPORT. Source: Authors

It must be concluded that unlike in the case of bivariate relationships, where an outlying or influential observation might usually be easily spotted on the two-dimensional scatterplot, in the case of the multivariate analyses the bivariate charts are no longer reliable. As will be evidenced further in the chapter, in such multivariate analyses, the commonly applied tools for detection of the univariate outliers (such as three-sigma rule) are often very weak in warning against the presence of outstanding observations.

2.2   Tools for Detection of Outlying Observations

2.2.1   Identifying Univariate Outliers

We will present two commonly applied tools for the detection of outliers in case of single variables:

- The two-sigma and three-sigma ranges.
- The "inter-quartile range rule."

The three-sigma rule is based on the following features of the normal distribution of a variable (Wheeler and Chambers 1992; Pukelsheim 1994):

- About 68,3% of observations of a variable fall into the range determined by its arithmetic mean ± one standard deviation (one-sigma).
- About 95,5% of observations of a variable fall into the range determined by its arithmetic mean ± two standard deviations (two-sigma).
- About 99,7% of observations of a variable fall into the range determined by its arithmetic mean ± three standard deviations (three-sigma).

According to the one-sigma rule, if the variable is normally distributed, then about 31,7% of observations of that variable lie outside the range determined by its arithmetic mean and one standard deviation (about 15,7% on the lower side and about 15,7% on the upper side). If so many observations fall outside the one-sigma range, then they cannot be deemed to be the outliers. Simply it is not reasonable to treat almost one-third of all observations as outstanding. Thus, the one-sigma is not very useful in detecting outliers.

According to the two-sigma rule, if the variable is normally distributed, then about 4,5% of observations of that variable lie outside the range determined by its arithmetic mean and two standard deviations (about 2,25% on the lower side and about 2,25% on the upper side). The observations with values reaching beyond the two-sigma range are considered outliers, because they are considered rare.

Finally, according to the three-sigma rule, if the variable is normally distributed, then only about 0,3% of its observations lie outside the range determined by its arithmetic mean and three standard deviations (about 0,15% on the lower side and about 0,15% on the upper side). Thus, the observations with values reaching beyond the three-sigma range are really exceptional: only one out of 666 observations have smaller value than the lower bracket of the range and only one out of 666 observations have higher value than the upper bracket. Clearly, such observations can be deemed to be outliers.

When applying the above rules in checking for the presence of the outlying observations, it must be remembered that the two-sigma range is more rigorous. As compared to the three-sigma, it treats much more observations as outliers. This brings some risk of picking as outliers some observations with values which actually are not that outstanding. In contrast, the three-sigma range is much wider. Thus, only very few observations fall outside that range. This, in turn, brings the risk of overlooking some outliers and including them into the set of observations which fall within the three-sigma range. In our opinion, it is safer to detect more rather than fewer potential outliers. Therefore, in the further part of the book we will favor and apply the two-sigma range for detecting outstanding observations.

The calculation of the two-sigma and the three-sigma rules is presented on Table 2.5.

As Table 2.5 shows, the two-sigma range correctly points to the observation in which the visual inspection of Fig. 2.1 suggested the outstanding value. In Fig. 2.1, there was one person who seemed to be distinctively tall, with the stature of 225 cm, which is above the upper frontier of the two-sigma range (212,38 cm). In that sample, there were not any persons who seemed uniquely short. Accordingly, all the observations exceed the lower frontier of the two-sigma range (139,82 cm). However, if we base our conclusions on the three-sigma range, then all the observations from Fig. 2.1 fall within that range. Thus, the relatively tall tenth person is not treated as an outlier under the three-sigma range.

The conclusions are more mixed in the case of the data from Fig. 2.2. The relatively tall person with the stature of 225 cm is again detected by the two-sigma range (with the upper frontier of 220,31 cm) and is overlooked by the three-sigma range (with the upper frontier of 244,21 cm). However, the relatively short person

Table 2.5 The two-sigma and three-sigma ranges applied to the data presented in Figs. 2.1 and 2.2

| | Stature (in cm) | |
Number of persons	Data from Fig. 2.1	Data from Fig. 2.2
1	162	126
2	163	163
3	165	165
4	170	170
5	171	171
6	174	174
7	175	175
8	178	178
9	178	178
10	225	225
Arithmetic mean (AM)	176,10	172,50
Standard deviation (SD)	18,14	23,90
AM + 2 × SD	212,38	220,31
AM − 2 × SD	139,82	124,69
AM + 3 × SD	230,51	244,21
AM − 3 × SD	121,69	100,79

Source: Authors

with the stature of 126 cm is overlooked by both ranges. This observation exceeds the lower frontier of both the two-sigma range (124,69 cm) as well as the three-sigma range (100,79 cm).

Most people would agree that the adult person with the stature of only 126 cm should be considered uniquely low. Although there are adults with such a stature, most of us do not even know anyone like this. However, in our sample of data, such person was not detected as an outstanding observation. Thus, the statistical outliers are not always consistent with the real-life outliers, if the two-sigma or three-sigma ranges are applied. This is the main drawback of this approach to outlier detection. The reason why such relatively short person was overlooked in the quest for outliers is simple: in searching for the outliers we applied the standard deviation, which itself is not immune to these very outliers. The presence of the outliers in the sample, particularly if they occur on both sides of a variable distribution, stretches the standard deviation. Thus, the more exceptional are the outliers, the wider are the two-sigma and three-sigma ranges. As a result, even some obvious outliers may fall within these ranges. In Table 2.5, we can notice that the standard deviation of the data from Fig. 2.2 is much wider than the equivalent value computed for the data from Fig. 2.1. Both samples differ by only one observation which is that very outlier. However, the presence of that low person in the sample has significantly stretched the standard deviation and as a result also the two-sigma and three-sigma ranges. The lower bound of the two-sigma range in the sample of data from Fig. 2.2 (equaling 124,69 cm) is lower by almost 11% from the respective bound computed for the data from Fig. 2.1 (equaling 139,82 cm).

To conclude, although the two-sigma and three-sigma ranges are commonly applied in the statistical analyses, they are not always reliable. These ranges are prone to some kind of paradox: they are applied for detecting outliers while they are themselves vulnerable to these very outliers. Therefore, for more credible picking of the outliers, we should search for some procedures which may be more immune to these outliers. One of the simplest approaches is known as the "inter-quartile range rule" (Tukey 1977). Its steps are presented in Table 2.6.

The computation of the "inter-quartile range rule" is presented in Table 2.7, on the basis of the same data as in Table 2.5. As can be seen, the quartiles computed for both samples are identical; thus also the inter-quartile ranges do not differ, unlike

Table 2.6 The procedure of "inter-quartile range rule"

Step 1	Compute the quartile 1 (denoted as Q_1) and quartile 3 (denoted as Q_3) in the investigated sample of data
Step 2	Compute the inter-quartile range (denoted as IQR) as follows: $IQR = Q_3 - Q_1$
Step 3	Determine the upper and lower bounds (denoted as UB and LB, respectively) as follows: $UB = Q_3 + 1,5 \times IQR$, $LB = Q_1 - 1,5 \times IQR$
Step 4	Treat any observations within the sample with the values above UB or below LB as outliers

Source: Authors

Table 2.7 The "inter-quartile range rule" applied to the data presented in Figs. 2.1 and 2.2

Number of persons	Stature (in cm)	
	Data from Fig. 2.1	Data from Fig. 2.2
1	162	126
2	163	163
3	165	165
4	170	170
5	171	171
6	174	174
7	175	175
8	178	178
9	178	178
10	225	225
Quartile 3 (Q_3)	177,25	177,25
Quartile 1 (Q_1)	166,25	166,25
Inter-quartile range (IQR)	11,00	11,00
Upper bound ($Q_3 + 1,5 \times IQR$)	193,75	193,75
Lower bound ($Q_1 - 1,5 \times IQR$)	149,75	149,75

Source: Authors

the standard deviations shown in Table 2.5. Moreover, the inter-quartile range is narrower than both standard deviations, because it is much more immune to the distorting impact of the outstanding observations (which tend to stretch the standard deviations). As Table 2.7 shows, in these datasets the inter-quartile range correctly picks all three observations in which case the visual inspection of Figs. 2.1 and 2.2 suggested the outstanding values. Particularly, the relatively short person with the stature of 126 cm now falls beyond the lower bound, as determined by the inter-quartile range (149,75 cm).

To conclude, as compared to the two-sigma or three-sigma ranges, here the "inter-quartile range rule" seems to constitute an approach which is more immune to distorting impact of outliers. Unfortunately, as will be illustrated later, also the "inter-quartile range rule" is not fully reliable and in some instances it is actually less credible than the two-sigma range. Furthermore, neither the two-sigma range nor the "inter-quartile range rule" are reliable in detecting multivariate outliers. We will show this on the basis of the data from Table 2.2 and Fig. 2.9. Table 2.8 reminds these data and provides the lower and upper bounds for both variables, as computed by the two-sigma range and the "inter-quartile range rule."

As Table 2.8 shows, the upper bound for the stature equals 183,21 cm, as determined by the two-sigma range, and 191 cm, as computed according to the "inter-quartile range rule." Meanwhile, the maximum stature within the sample is 180 cm, which means that none of the observations exceeds any of the two alternative upper bounds. Similarly, the upper bound for the weight equals 82,07 kg, as determined by the two-sigma range, and 90 kg, as computed according to the "inter-quartile range rule," while the maximum weight within the sample is 80 kg. We can conclude, therefore, that none of the observations of the stature and the weight reaches beyond any of the respective upper frontiers. The lower bounds for the stature are 158,70 cm (the two-sigma range) or 151 cm (the "inter-quartile range rule"), while the minimum stature within the sample is 161 cm. The lower bound for the weight is 58,31 kg (the two-sigma range) or 50 kg (the "inter-quartile range rule"), while the minimum weight within the sample is 61 kg. Thus, we can conclude that none of the observations of the stature and the weight reaches beyond any of the respective lower frontiers.

According to the evidence presented in Table 2.8, neither the two-sigma range nor the "inter-quartile range rule" was able to detect the bivariate outlier, which is the last observation in the sample. Figures 2.9 and 2.10 showed how dramatically the inclusion of that last observation impacts the parameters of the regression of the weight against the stature. Overlooking the outstanding nature of that outlier results in obtaining unreliable regression results. However, as we could see, none of the tools for identifying the univariate outliers was helpful here. Therefore, in bivariate and particularly multivariate analyses, we have to employ some other approaches for detecting the outliers. As we will show, however, it does not mean that the tools discussed so far will not be useful.

Table 2.8 The stature (in cm) and the weight (in kg) of a group of persons (the same data as in Table 2.2) and the upper and lower bounds, as computed by the two-sigma range and the "inter-quartile range rule"

Number of persons	Stature (in cm)	Weight (in kg)
1	161	61
2	162	62
3	163	63
4	164	64
5	165	65
6	166	66
7	167	67
8	168	68
9	169	69
10	170	70
11	171	71
12	172	72
13	173	73
14	174	74
15	175	75
16	176	76
17	177	77
18	178	78
19	179	79
20	180	80
21	180	64
Arithmetic mean (AM)	170,95	70,19
Standard deviation (SD)	6,13	5,94
Upper bound (AM + 2 × SD)	183,21	82,07
Lower bound (AM − 2 × SD)	158,70	58,31
Quartile 3 (Q_3)	176	75
Quartile 1 (Q_1)	166	65
Inter-quartile range (IQR)	10	10
Upper bound ($Q_3 + 1,5 \times$ IQR)	191	90
Lower bound ($Q_1 - 1,5 \times$ IQR)	151	50
Max in the sample	180	80
Min in the sample	161	61

Source: Authors

2.2.2 Identifying Multivariate Outliers

We will present two alternative tools for the detection of outliers in the case of multivariate relationships:

- The cross-validation.
- The analysis of the residuals from the regression.

Table 2.9 The procedure of cross-validation

Step 1	For the set of data with n observations, numbered $n_1, n_2, \ldots, n_n$, estimate n regressions of a dependent variable against its explanatory variables, each of them on the basis of $n - 1$ observations: • In the first regression, omit the observation numbered n_1 and estimate it using the observations numbered $n_2, n_3, \ldots, n_n$ • In the second regression, omit the observation numbered n_2 and estimate it using the observations numbered $n_1, n_3, \ldots, n_n$ • . . . . • In the n-th regression, omit the observation numbered n_n and estimate it using the observations numbered $n_1, n_2, \ldots, n_{n-1}$
Step 2	On the basis of the regressions estimated in Step 1, compute the predictions of the dependent variable. For a given observation, predict the dependent variable on the basis of the regression in which this observation was omitted: • Predict the observation of dependent variable numbered n_1 using the regression estimated on the basis of observations numbered $n_2, n_3, \ldots, n_n$ • Predict the observation of dependent variable numbered n_2 using the regression estimated on the basis of observations numbered $n_1, n_3, \ldots, n_n$ • . . . . • Predict the observation of dependent variable numbered n_n using the regression estimated on the basis of observations numbered $n_1, n_2, \ldots, n_{n-1}$
Step 3	For n forecasts of a dependent variable obtained in Step 2, compute n forecast errors (FE_{n_k}) as follows: $$FE_{n_k} = Y_{n_k} - \hat{Y}_{n_k},$$ where: Y_{n_k}—actual value of the dependent variable in observation numbered n_k, $\hat{Y}_{n_k}$—predicted value of the dependent variable in observation numbered n_k
Step 4	For the series of n forecast errors obtained in Step 3, check whether any of these errors is an outlying observation, by means of the two-sigma range or the "inter-quartile range rule." Treat any outlying forecast error as the multivariate outlier

Source: Authors

Cross-validation is a procedure which is commonly applied in testing the forecasting accuracy of an estimated model (Geisser 1993). It may also be applied in searching for outliers, according to the procedure presented in Table 2.9.

We will illustrate the computation of the cross-validation on the basis of data from Table 2.3. As we remember, the regression of WEIGHT against STATURE, CALORIES, and SPORT resulted in obtaining an evidently outlying residual for the last observation (which was clearly visible in Figs. 2.11 and 2.15). However, neither of the three bivariate relationships between WEIGHT and its individual drivers (plotted in Figs. 2.12, 2.13, and 2.14) suggested any outstanding observation. Furthermore, as is evidenced in Table 2.10, neither the two-sigma range nor the "inter-quartile range rule" applied to the individual variables of that regression could warn us against this multivariate outlier (the last observation). None of the maximum values of the four individual variables exceeds its respective upper bound, both from the two-sigma range as well as from the "inter-quartile range rule." Likewise, none of the minimum values of these variables is lower than its respective lower bound, as determined by the two alternative rules. Thus, neither

Table 2.10 The two-sigma range and the "inter-quartile range rule" applied to the individual variables of the regression of WEIGHT against STATURE, CALORIES, and SPORT (on the basis of data from Table 2.3)

	WEIGHT (actual weight in kg)	STATURE (stature in cm)	CALORIES (average daily calories consumption in kcal)	SPORT (average weekly sporting activity in hours)
Arithmetic mean (AM)	75,69	170,95	2.561,90	4,57
Standard deviation (SD)	11,39	6,13	347,10	3,94
Upper bound (AM + 2 × SD)	98,47	183,21	3.256,10	12,45
Lower bound (AM − 2 × SD)	52,91	158,69	1.867,70	−3,31
Quartile 3 (Q_3)	83,15	176,00	2.800,00	6,00
Quartile 1 (Q_1)	68,80	166,00	2.400,00	2,00
Inter-quartile range (IQR)	14,35	10,00	400,00	4,00
Upper bound ($Q_3 + 1.5 \times$ IQR)	104,68	191,00	3.400,00	12,00
Lower bound ($Q_1 - 1.5 \times$ IQR)	47,28	151,00	1.800,00	−4,00
Max in the sample (MAX)	95,7	180,00	3.200,00	12,00
Min in the sample (MIN)	56,5	161,00	1.900,00	0,00
Is MAX larger than the upper bound from the two-sigma range?	NO	NO	NO	NO
Is MIN lower than the lower bound from the two-sigma range?	NO	NO	NO	NO
Is MAX larger than the upper bound from the IQR rule?	NO	NO	NO	NO
Is MIN lower than the lower bound from the IQR rule?	NO	NO	NO	NO

Source: Authors

the two-sigma range nor the "inter-quartile range rule," when applied to the individual variables of the regression of WEIGHT, suggests any outstanding observation.

Table 2.11 presents the computations of forecast errors of WEIGHT (Steps 1–3 discussed in Table 2.9). Four columns denoted as "Regression parameters" contain the parameters of the individual regressions of WEIGHT against their three explanatory variables (Step 1). For example, the parameters for the first observation stem

Table 2.11 The computation of forecast errors of WEIGHT for detection of multivariate outliers by means of the cross-validation (on the basis of data from Table 2.3)

Number of persons	Actual WEIGHT	Regression parameters (intercept and slope parameters)				Predicted WEIGHT[a]	Forecast error[b]
		Intercept	STATURE	CALORIES	SPORT		
1	83,1	−43,772	0,379	0,024	−1,643	77,68	5,37
2	89,1	−43,227	0,384	0,024	−1,615	82,58	6,52
3	95,7	−42,841	0,398	0,022	−1,602	87,69	7,96
4	58,2	−25,695	0,285	0,024	−1,653	61,57	−3,37
5	62,3	−26,336	0,282	0,024	−1,639	65,55	−3,30
6	78,8	−34,579	0,325	0,024	−1,684	77,59	1,21
7	72,9	−32,992	0,316	0,024	−1,686	73,16	−0,31
8	60,4	−30,287	0,314	0,023	−1,691	63,74	−3,34
9	74,5	−33,314	0,316	0,024	−1,678	73,88	0,57
10	56,5	−16,931	0,224	0,024	−1,369	66,70	−10,20
11	84,6	−33,114	0,312	0,025	−1,676	85,75	−1,20
12	80,1	−33,663	0,318	0,024	−1,687	81,18	−1,08
13	83,2	−33,832	0,319	0,024	−1,688	83,93	−0,78
14	79,2	−34,515	0,325	0,024	−1,697	80,33	−1,13
15	56,8	−36,684	0,366	0,023	−1,638	64,81	−8,06
16	68,8	−36,168	0,316	0,025	−1,496	77,74	−8,94
17	72,9	−39,524	0,350	0,025	−1,655	78,27	−5,42
18	80,9	−36,090	0,336	0,024	−1,713	82,00	−1,10
19	79,0	−40,194	0,361	0,024	−1,735	81,68	−2,73
20	86,0	−39,634	0,351	0,025	−1,720	88,31	−2,31
21	87,0	0,000	0,050	0,030	−2,500	51,00	36,00

Source: Authors
[a] Prediction of a k-th observation of WEIGHT from the regression of WEIGHT against STATURE, CALORIES, and SPORT, estimated after omitting that k-th observation
[b] Actual weight − predicted weight

from the regression estimated on the basis of the 20 observations numbered 2, 3, . . ., 21, the parameters for the second observation stem from the regression estimated on the basis of the 20 observations numbered 1, 3, . . ., 21, and the parameters for the last observation stem from the regression estimated on the basis of the 20 observations numbered 1, 2, . . ., 20. Predicted weight (Step 2) was computed on the basis of the respective parameters shown in the columns denoted as "Regression parameters" and the actual values of the explanatory variables for a given observation (from Table 2.3). Finally, forecast errors (Step 3) are the differences between the actual and predicted weight.

Now we can proceed to checking whether any of the forecast errors presented in the last column of Table 2.11 have outlying values, by means of either the two-sigma range or the "inter-quartile range rule." The results of this analysis are presented in Table 2.12.

As can be seen, now the two-sigma range seems to be more accurate in picking outliers than the "inter-quartile range rule." According to the two-sigma range, there is only one outstanding observation in the sample: the last one. This is consistent with our earlier findings, visualized in Figs. 2.11 and 2.15. In contrast, the "inter-quartile range rule," while correctly pointing to the last observation, picked also the other three observations as the outliers. This seems not to be supported by the data presented in Figs. 2.11 and 2.15. Thus, in this particular case, the two-sigma range seems to constitute more credible tool for detecting outliers. However, the example presented earlier in Table 2.7 showed that in that case the "inter-quartile range rule" was more accurate. Therefore, unfortunately none of these two alternative tools unequivocally outperforms the another one and their relative accuracy varies in different datasets.

Now we will illustrate the alternative and less time-consuming approach for detecting the multivariate outliers, which is the analysis of the residuals from the regression. Its procedure is presented in Table 2.13.

We will illustrate the analysis of the residuals from the regression again on the basis of data from Table 2.3. The next to last column of Table 2.3 contains the fitted values of WEIGHT obtained from the regression (estimated on the basis of all 21 observations) of WEIGHT against its three drivers. The last column of Table 2.3 presents the residuals of that regression. As has been illustrated in Fig. 2.11, the last residual from that regression seems to have an outstanding value. Let's now check whether this presumption is true by means of the two-sigma range and the "inter-quartile range rule" applied to the residuals from the last column of Table 2.3. The results of this analysis are presented in Table 2.14.

As in the case of the cross-validation, the two-sigma range seems to be more accurate in picking the outliers than the "inter-quartile range rule." According to the two-sigma range, there is only one outstanding observation in the sample: the last one. This is consistent with our earlier findings, visualized in Figs. 2.11 and 2.15, as well as our findings obtained for the cross-validation (presented in Table 2.12). In contrast, the "inter-quartile range rule," while correctly pointing to the last observation, picked also the other three observations as the outliers (the same ones as in the case of the cross-validation). Thus, in the case of this particular dataset, the

Table 2.12 The identification of outstanding forecast errors of WEIGHT by means of the two-sigma range and the "inter-quartile range rule" (on the basis of data from the last column of Table 2.11)

Number of persons	Forecast error of WEIGHT[a]	Is the observation an outlier according to the two-sigma range?	Is the observation an outlier according to the "inter-quartile range rule"?
1	5,37		
2	6,52		YES
3	7,96		YES
4	−3,37		
5	−3,30		
6	1,21		
7	−0,31		
8	−3,34		
9	0,57		
10	−10,20		YES
11	−1,20		
12	−1,08		
13	−0,78		
14	−1,13		
15	−8,06		
16	−8,94		
17	−5,42		
18	−1,10		
19	−2,73		
20	−2,31		
21	36,00	YES	YES
Arithmetic mean (AM)	0,21		
Standard deviation (SD)	9,39		
Upper bound ($AM + 2 \times SD$)	18,99		
Lower bound ($AM - 2 \times SD$)	−18,57		
Quartile 3 (Q_3)			0,57
Quartile 1 (Q_1)			−3,34
Inter-quartile range (IQR)			3,91
Upper bound ($Q_3 + 1,5 \times IQR$)			6,44
Lower bound ($Q_1 - 1,5 \times IQR$)			−9,21

Source: Authors

[a]Data from the last column of Table 2.9

Table 2.13 The procedure of analysis of the residuals from the regression

Step 1	Estimate the regression of dependent variable against its explanatory variables and compute the residuals from that regression
Step 2	For the series of residuals obtained in Step 1 check whether any of these residuals is an outlying observation, by means of the two-sigma range or the "inter-quartile range rule." Treat any outlying residual as the multivariate outlier

Source: Authors

two-sigma range seems to be a more credible tool for detecting outliers, both when used for the cross-validation as well as for the analysis of the residuals from the regression.

As is presented in Tables 2.12 and 2.14, both the cross-validation as well as the analysis of the regression residuals correctly picked the last observation as the multivariate outlier, if the two-sigma range was applied (instead of the "inter-quartile range rule"). Both tools were successful in pointing to the last observation, even though neither the two-sigma range nor the "inter-quartile range rule" applied to the individual variables in the regression signaled any outstanding observations. However, it does not mean that the cross-validation and the analysis of the regression residuals are perfect in warning against the outliers in regression analysis. As we will illustrate below, there are some situations in which the analysis of the observations of individual variables (by means of the two-sigma range nor the "inter-quartile range rule") correctly identifies the outliers while both the cross-validation and the analysis of the regression residuals fail to do so. This tends to happen when there are two or more heavily influential observations which lie close to each other.

Let's turn back to the data presented in Table 2.1. But now imagine that we have one additional influential observation (11th person in the sample). This set of data in shown in the second and third columns of the upper part of Table 2.15. The relationship between both variables within that sample is visualized in Fig. 2.16. The upper part of Table 2.15 contains also the obtained forecast errors for the cross-validation analysis as well as the residuals from the regression of weight against stature. The lower part of the table provides the obtained upper and lower bounds of both variables (stature and weight) as well as the upper and lower bounds of the forecast errors and residuals of weight, as obtained by means of the two-sigma range and the "inter-quartile range rule."

According to Table 2.15, the forecast error for neither 10th nor 11th observation exceeds the respective upper or lower bounds, as determined by the two-sigma range or the "inter-quartile range rule." Likewise, none of the residuals from the regression of weight against stature, obtained for the last two observations, reaches beyond the ranges determined by the two-sigma or the "inter-quartile range rule." Actually, the values of these forecast errors and residuals lie rather far from their respective bounds. Thus, neither the cross-validation nor the analysis of the regression residuals would warn against the distorting effect of these two heavily influential observations. In contrast, the analysis of the observations of individual variables produces some warning signals. Both the stature and the weight of the tenth and the eleventh person exceed their respective upper bounds obtained from

Table 2.14 The identification of outstanding residuals from the regression of WEIGHT against its three drivers, by means of the two-sigma range and the "inter-quartile range rule" (on the basis of data from the last column of Table 2.3)

Number of persons	Residual[a]	Is the observation an outlier according to the two-sigma range?	Is the observation an outlier according to the "inter-quartile range rule"?
1	4,32		
2	5,18		YES
3	5,79		YES
4	−2,56		
5	−2,76		
6	1,11		
7	−0,29		
8	−2,57		
9	0,49		
10	−7,29		YES
11	−0,92		
12	−1,00		
13	−0,70		
14	−1,04		
15	−6,57		
16	−6,64		
17	−4,78		
18	−0,79		
19	−2,12		
20	−1,76		
21	24,90	YES	YES
Arithmetic mean (AM)	0,00		
Standard deviation (SD)	6,70		
Upper bound ($AM + 2 \times SD$)	13,40		
Lower bound ($AM - 2 \times SD$)	−13,40		
Quartile 3 (Q_3)			0,49
Quartile 1 (Q_1)			−2,57
Inter-quartile range (IQR)			3,06
Upper bound ($Q_3 + 1,5 \times IQR$)			5,08
Lower bound ($Q_1 - 1,5 \times IQR$)			−7,16

Source: Authors

[a]Data from the last column of Table 2.3

Table 2.15 The stature (in cm) and the weight (in kg) of 11 persons with two influential observations on the upper side

Number of persons	Stature (in cm)	Weight (in kg)	Forecast error of Weight[a]	Residual[b]
1	162	65	4,16	3,50
2	163	60	−3,25	−2,76
3	165	75	11,28	9,72
4	170	67	−5,16	−4,58
5	171	72	−0,94	−0,84
6	174	70	−7,35	−6,62
7	175	80	2,35	2,12
8	178	76	−6,23	−5,66
9	178	85	3,68	3,34
10	225	140	−1,72	−0,87
11	223	141	4,88	2,65
Arithmetic mean (AM)	180,36	84,64	0,15	0,00
Standard deviation (SD)	22,27	28,48	5,67	4,85
Upper bound (AM + 2 × SD)	224,90	141,60	11,49	9,70
Lower bound (AM − 2 × SD)	135,82	27,68	−11,19	−9,70
Quartile 3 (Q_3)	178,00	82,50	3,92	3,00
Quartile 1 (Q_1)	167,50	68,50	−4,21	−3,67
Inter-quartile range (IQR)	10,50	14,00	8,13	6,67
Upper bound ($Q_3 + 1,5 \times$ IQR)	193,75	103,50	16,12	13,01
Lower bound ($Q_1 − 1,5 \times$ IQR)	151,75	47,50	−16,41	−13,68

Source: Authors

[a]Actual weight − predicted weight (where prediction of a k-th observation of Weight is obtained from the regression of Weight against Stature, estimated after omitting that k-th observation)

[b]From the regression of Weight against Stature, of the following form: Weight $= -142,592 + 1,260 \times$ Stature

the "inter-quartile range rule." Less clear findings are obtained from their respective two-sigma ranges, where the tenth and the eleventh observations of both variables either lie slightly below the upper bound or exceed it only marginally.

As was illustrated in the above example, in some circumstances the analysis of the univariate outliers turns out to be more credible than any of the two methods for identifying multivariate outliers. This is common for heavily influential observations. However, as evidenced by some of the previous examples, in other situations the cross-validation and the analysis of regression residuals are capable of correctly picking the outliers which are overlooked by the analysis of the single variables. Therefore, the analysis of the observations of individual regression variables and the search for the multivariate outliers should be treated as complementary, not supplementary tools of outlier detection. In the following section, we will discuss the recommended procedure for the outliers detection in the regression analysis.

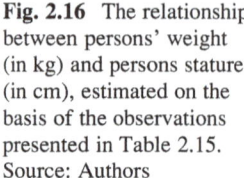

Fig. 2.16 The relationship between persons' weight (in kg) and persons stature (in cm), estimated on the basis of the observations presented in Table 2.15. Source: Authors

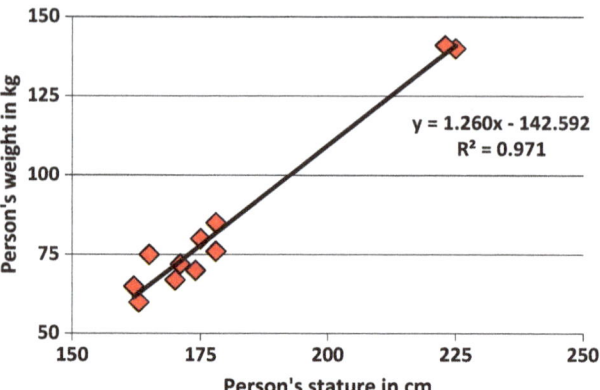

2.3 Recommended Procedure for Detection of Outlying and Influential Observations

As the examples presented in the preceding sections showed, none of the discussed approaches for outlier detection unequivocally outperforms the others. In some circumstances, the analysis of the single variables was able to identify outstanding observations which were overlooked by the multivariate analysis, while in other situations the cross-validation and the analysis of regression residuals were capable of picking multivariate outliers when there were no suspicious univariate observations. Thus, in regression analysis we recommend using the procedure combining the two approaches for the outliers detection: the one based on the observations of individual variables and the another one focused on the multivariate analysis.

During the analysis of the observations of single variables as well as when searching for the multivariate outliers, the upper and lower bounds for the observations must be determined. In the preceding sections, we offered two approaches for setting these bounds: the two-sigma range (which is based on the arithmetic mean and the standard deviation) and the "inter-quartile range rule" (which is based on the quartiles). The question arises therefore: which of these two approaches is more reliable and should be followed? Unfortunately, there is not a clear answer to that question. As was evidenced before, in some circumstances the two-sigma range is more accurate while at other occasions the "inter-quartile range rule" turns out to generate more credible findings. Therefore, we recommend combining both approaches while keeping in mind that they may provide inconsistent suggestions. In cases when the two approaches contradict each other, the common sense must be applied and final classification of a given observation as the outlier or not should take into consideration the subjective factors and potential risks of including the outliers within the sample. Alternatively, the following rules could be followed:

Table 2.16 The recommended procedure of detecting the outliers for regression analysis

Step 1	Identify the outlying univariate observations of the dependent variable as well as all the candidate explanatory variables, by means of the "two-sigma range" and the "inter-quartile range rule"
Step 2	Regress the dependent variable against all the candidate explanatory variables and compute the residuals from such a regression
Step 3	For the series of residuals obtained in Step 2, identify the outlying multivariate observations, by means of the "two-sigma range" or the "inter-quartile range rule." Treat any outlying residual as the multivariate outlier

Source: Authors

- If neither the two-sigma range nor the "inter-quartile range rule" suggests an observation to be an outlier, this is not an outlier.
- If both the two-sigma range and the "inter-quartile range rule" suggest an observation to be an outlier, this is an outlier.
- If one of the approaches picks the observation as an outlier while under the alternative approach the observation lies within the range determined by the respective upper and lower bound and far from these bounds, this is not an outlier.
- If one of the approaches picks the observation as an outlier while under the alternative approach the observation lies within the range determined by the respective upper and lower bound but close to one of these bounds, this is an outlier.

Clearly, stating whether the observation lies far from or close to the upper or lower bound requires a subjective judgment. However, this is an inevitable feature of any practical statistical analysis.

The recommended procedure for identifying outliers for regression analysis is presented in Table 2.16. Although in the previous sections we discussed two approaches for identifying multivariate outliers, we prefer applying the analysis of the regression residuals, instead of the cross-validation. This is so because both approaches seem to have comparable accuracy, while the analysis of the regression residuals is less time-consuming.

2.4 Dealing with Detected Outlying and Influential Observations

After detecting the outstanding observations within the dataset, some treatment of these outliers must be chosen. Basically, in the regression analysis the four general options are available:

- Ignore the outlier and do nothing with it.
- Remove the outlier from the sample.
- Replace the actual observation of an outlier by some alternative observation.
- Capture the impact of the outlier by means of one or more dummy variables.

Ignoring the outlier seems to constitute a reasonable option in cases when the outlying observation does not deviate that much from its respective upper or lower bound. In such cases, leaving its original observation in the sample does not bring about any significant distortion of the regression analysis.

Another option is to remove the outlier from the sample. This is usually quite easy in the case of the cross-sectional data (where individual observations are not related to each other), but it is often more complicated in the case of the time-series models (where the order of observations is fixed and related to the time). Particularly, in the time-series models with lagged explanatory variables, removing the outlier occurring in only one period entails removal of several adjacent observations. This problem will be illustrated with the numerical example in Chap. 5.

If an analyst does not want to ignore or remove the outlier, than the distorting impact of that outstanding observation should be neutralized in some way. One of the available options is to replace the actual outlying value by some artificial value which is not that distinctive. Among the common approaches are:

- Replacing the outlying observation by its respective upper or lower bound (e.g., as determined by the two-sigma range).
- Replacing the outlying observation by the arithmetic average of its neighboring observations.
- Replacing the outlying observation by its smoothed value, obtained from, e.g., trend of that variable or a regression model, in which that variable is a dependent variable.

When the outlying observation is replaced by its respective upper or lower bound, then its value remains distinctively high or low, respectively, but to a lesser extent than its original value. Therefore, the outlier with the outstandingly high value keeps the relatively high value (as compared to other observations), but has weaker and less distorting impact on the results of the statistical analysis. Likewise, the outlier with the outstandingly low value takes the another relatively low but less distorting value. As a result, the outliers keep on having uniquely high or low values, but their distorting effect on regression results is mitigated.

Replacing the outlying observation by the arithmetic average of its neighboring observations is usually applied for the time series which do not show any visible trends. In such cases, the outlier is brought about by some one-off event or even a measurement error. The arbitrary decision here is the number of the neighboring observations from which the arithmetic average is calculated.

Replacing the outlying observation by its smoothed value, obtained from the linear or nonlinear trend function of a given variable, is usually applied for the time series which show discernible trends. In such cases, the outlier is removed and replaced by the fitted value of a respective observation from the estimated trend line. In contrast, the trend functions cannot be estimated for the cross-sectional data. In such cases, an auxiliary regression model may be estimated, in which the variable with the outlier constitutes a dependent variable while some other variables are used as explanatory variables. Then the outlier is replaced by its fitted value obtained from such an auxiliary regression.

Finally, the another way of dealing with outliers is to capture their effect by means of one or more dummy variables. This will be discussed in detail and exemplified in Chaps. 3 and 5.

References

Geisser S (1993) Predictive inference. An introduction. Chapman & Hall, London

Makridakis S, Weelwright SC, Hyndman RJ (1998) Forecasting. Methods and applications. Wiley, Hoboken

Pukelsheim F (1994) The three sigma rule. Am Stat 48:88–91

Tukey JW (1977) Exploratory data analysis. Addison-Wesley, Boston

Wheeler DJ, Chambers DS (1992) Understanding statistical process control. SPC Press, Knoxville

Chapter 3
Basic Procedure for Multiple Regression Model Building

3.1 Introduction

In this chapter, we will illustrate the step-by-step procedure for the estimation of an econometric model. Starting from the raw set of hypothetical data (concerning a fictitious company), we will guide the reader through the following phases of the model building process:

- The preliminary specification of the model.
- The detection of potential outliers in the data set.
- The selection of explanatory variables from the set of candidate variables (by means of two alternative procedures, that is, "general to specific modeling" and "stepwise regression").
- The tentative interpretation of the obtained regression structural parameters.

We will illustrate each phase on the basis of a hypothetical company operating in the fast-food business. The company developed a chain of points of sales of hamburgers (and supplementary items such as soft drinks, tea, coffee, and some confectionery). The company's operations at the end of 2012 are run in 150 sales points, up from about 100 at the beginning of 2008, and the company's fundamental competitive advantage is the location of its sales points, in close proximity to the central points of the business districts in the middle-sized cities. However, the company was managed so far with a good deal of the management intuition instead of any formal and structured business analysis tools, and the managing board has recently become concerned about the declining and highly variable profit margins.

© Springer International Publishing AG 2018
J. Welc, P.J.R. Esquerdo, *Applied Regression Analysis for Business*,
https://doi.org/10.1007/978-3-319-71156-0_3

3.2   Preliminary Specification of the Model

The management of our company is concerned about quite substantial instability of the company's quarterly gross profit margin. Quarterly gross profit margin is computed as the ratio of a quarterly gross profit on sales to quarterly net sales. It informs how many cents are earned from every dollar of net sales, on average. The data regarding the quarterly gross profit of this company, in the 5-year period, are presented on Table 3.1.

The data from the last two columns of the above table are visualized in Fig. 3.1. Indeed, as we can see, the company's gross margin on sales seems to be quite variable. This variability can definitely increase the company's operating risks, because the more variable and unpredictable the margin is, the more difficult it is to budget the company's future financial results. The arithmetic average of the quarterly margin in the analyzed 5-year period equaled 25%, with the standard deviation of 2,7% points. This gives the variation of about ±10,8% around the arithmetic average. Moreover, the gross margin on sales is not only quite variable, but it also seems to be quite "noisy" (or "random"), with visible jumps from quarter to quarter. Finally, the management is not only concerned about this noisiness of the gross margin, but also about its declining trend observed in the last four quarters.

This situation has motivated the company's management to construct an econometric model which would help capture the statistical relationships between the company's quarterly costs of goods sold and the potential drivers of these costs. On the basis of the experience of the sales force members, the management team agreed that the company quarterly net sales are predictable with a reasonable degree of accuracy, so the cost of goods sold is the problematic variable, in terms of their predictability and manageability.

In our model, the dependent variable is quarterly cost of goods sold (in thousands of USD), denoted as $COGS_t$, where t denotes a particular quarter. As the candidate explanatory variables, the management proposed the following:

- **NS_t—quarterly net sales (in thousands of USD)**
  This variable is the proxy for the scope of the company's operations. It is assumed to have strong positive causal relationship with cost of goods sold, because the more You sell, the more costs You need to incur (e.g., for raw materials used, employees' remuneration, etc.).
- **PRM_t—quarterly wholesale price of raw materials (beef) per package (in USD)**
  Management decided to include this variable on the ground that the purchase (wholesale) price of the raw meat was cyclical and variable in the last couple of years. It is confirmed by the chart below, which shows clearly that this was the case. This variable is assumed to have a positive relationship with costs of goods sold (Fig. 3.2).

Table 3.1 Quarterly net sales, cost of goods sold, gross profit on sales, and gross margin on sales of a company

Period	Quarterly net sales in thousands USD	Quarterly costs of goods sold in thousands USD	Quarterly gross profit on sales	Gross margin on sales (quarterly) (%)	Gross margin on sales in the last four quarters (%)[a]
	1	2	3 (1 − 2)	4 (3/1)	5
I q. 2008	11.200	8.700	2.500	22,3	N/A
II q. 2008	13.400	10.525	2.875	21,5	N/A
III q. 2008	12.700	10.235	2.465	19,4	N/A
IV q. 2008	15.000	11.750	3.250	21,7	21,2
I q. 2009	14.500	11.145	3.355	23,1	21,5
II q. 2009	16.900	12.235	4.665	27,6	23,2
III q. 2009	16.400	11.925	4.475	27,3	25,1
IV q. 2009	16.000	11.420	4.580	28,6	26,8
I q. 2010	17.500	12.530	4.970	28,4	28,0
II q. 2010	15.800	12.240	3.560	22,5	26,8
III q. 2010	19.000	14.155	4.845	25,5	26,3
IV q. 2010	20.300	15.165	5.135	25,3	25,5
I q. 2011	18.500	14.275	4.225	22,8	24,1
II q. 2011	19.600	14.575	5.025	25,6	24,8
III q. 2011	21.900	15.765	6.135	28,0	25,6
IV q. 2011	21.000	15.050	5.950	28,3	26,3
I q. 2012	20.000	14.655	5.345	26,7	27,2
II q. 2012	19.000	14.330	4.670	24,6	27,0
III q. 2012	20.000	14.800	5.200	26,0	26,5
IV q. 2012	19.000	14.295	4.705	24,8	25,5

Source: Authors
[a]Sum of the gross profit on sales in the last four quarters divided by the sum of the net sales in the last four quarters

- **APOS,—total number of all points of sale (POS) at the end of the quarter**
  In the course of the analyzed 5-year period, the company opened almost 50 new points of sales, increasing their total number from 102 at the end of the first quarter of 2008 to 150 at the end of 2011. Management hypothesizes that part of the erratic behavior of the quarterly gross margin can stem from some inefficiencies in managing the rapidly growing chain of points of sale. For example, there may be some problems with coordinating the management of supplies to the individual points of sales, when their number is increasing very fast. At this moment, the management does not have any hypotheses about the direction of the relationship between this variable and the costs of goods sold.

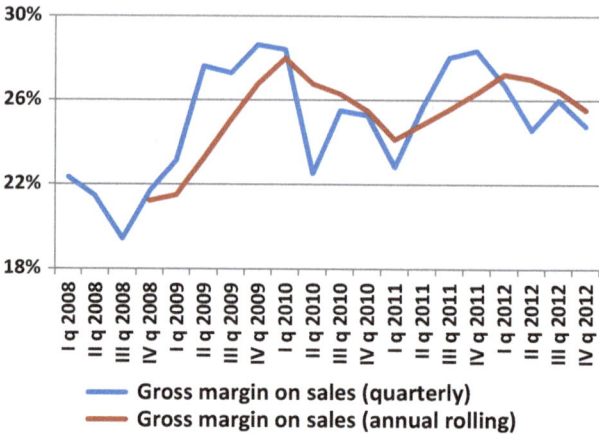

Fig. 3.1 Quarterly and rolling gross margin on sales (based on the data from the Table 3.1).
Source: Authors

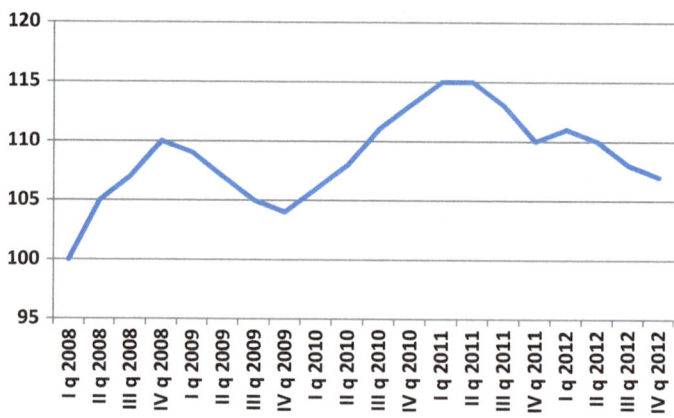

Fig. 3.2 Quarterly average price of raw meat in USD per package (fictitious data). Source:
Authors

- **NPOS$_t$—the number of new points of sale (POS) opened in a given quarter**
  Any new opening can temporarily change the relationships between sales and
  costs. For example, newly opened POS can already incur major operating
  expenses, and even some additional expenditures, such as for intensive promo-
  tion, while its base of customers is only being created, so the sales can be still
  relatively low in the first few months of operation. Management hypothesizes
  that this variable is positively related to the costs of goods sold, meaning that the
  more new POS openings, the higher the costs, due to the initial start-up costs at
  the limited consumer base.

- **NOE$_t$—total number of employees at the end of the quarter**

  The impact of the salaries of direct sales staff at the points of sales is probably already captured by the APOS$_t$, because these two variables should be strongly positively correlated. The more points of sales, the more people employed by the company. The salaries of the other employees are not included in costs of goods sold since they make up general and administrative expenses. However, management decided to include this variable in the preliminary set of candidate explanatory variables, on the ground that some of the remuneration issues, including bonuses or other pay increases, can be overlooked by the APOS$_t$. The variable is assumed to have a positive relationship with cost of goods sold.

- **SOH$_t$—share of the revenues from sales of hamburgers in total net revenues in the quarter**

  In its points of sales, the company offers and sells not only hamburgers but also supplementary refreshments like soft drinks, snacks, tea and coffee, and some confectionary. Because the gross margins on these items differ, the total company gross margin shifts in line with the changes of the sales breakdown. In the winter months, the sales volume of cold soft drinks declines, but the sale of tea and coffee increases. In contrast, in the summertime the sales of soft drinks jump up and some people even purchase only soft drinks, without any hamburgers. In light of the differences in these products gross margins, it is justified to include in our econometric model the candidate explanatory variable reflecting the impact of changing sales breakdown on company total cost of goods sold. Gross margins from the sales of items other than hamburgers are much lower than the margins on hamburgers themselves, because these supplementary items are offered only to boost the sales of hamburgers, thanks to the more comprehensive offer. Management hypothesizes that this variable is negatively related to the cost of goods sold, because the higher the share of hamburgers in total quarterly revenues, the higher the total margin, and as a result the lower the total cost of goods sold.

After completing the set of these six candidate explanatory variables, the company's analysts gathered all the statistical data required for the further analysis. These data are presented in Table 3.2.

3.3 Detection of Potential Outliers in the Dataset

The next step in our procedure of econometric modeling is to check for the presence of any outliers in our dataset. In order to do this, we will apply three alternative tools:

- Compare the medians of all variables with their arithmetic means (keeping in mind that this should serve only as a preliminary test).
- Apply the "three-sigma rule" to all the individual variables (keeping in mind that this also is sometimes unreliable).

Table 3.2 The statistical data for all the observations of the dependent variable and all the candidate explanatory variables

Period	Dependent variable COGS$_t$ Quarterly cost of goods sold in thousands USD	Candidate explanatory variables					
		NS$_t$ Quarterly net sales in thousands USD	PRM$_t$ Wholesale price of basic raw materials in USD/pack.	APOS$_t$ Total number of all POS at the end of the quarter	NPOS$_t$ Number of the new POS opened in the quarter	NOE$_t$ Total number of employees at the end of the quarter	SOH$_t$ Share of hamburgers in the total net sales' revenues (%)
I q. 2008	8.700	11.200	100	102	2	123	74,5
II q. 2008	10.525	13.400	105	104	2	126	78,0
III q. 2008	10.235	12.700	107	107	3	130	81,0
IV q. 2008	11.750	15.000	110	110	3	134	76,0
I q. 2009	11.145	14.500	109	115	5	140	75,0
II q. 2009	12.235	16.900	107	119	4	145	77,5
III q. 2009	11.925	16.400	105	124	5	151	79,0
IV q. 2009	11.420	16.000	104	130	6	158	74,5
I q. 2010	12.530	17.500	106	135	5	164	77,0
II q. 2010	12.240	15.800	108	138	3	168	78,5
III q. 2010	14.155	19.000	111	140	2	171	82,0
IV q. 2010	15.165	20.300	113	143	3	175	77,5
I q. 2011	14.275	18.500	115	145	2	178	75,0
II q. 2011	14.575	19.600	115	146	1	180	78,0
III q. 2011	15.765	21.900	113	149	3	184	78,0
IV q. 2011	15.050	21.000	110	150	1	186	73,0
I q. 2012	14.655	20.000	111	150	0	187	74,0
II q. 2012	14.330	19.000	110	150	0	188	80,0
III q. 2012	14.800	20.000	108	150	0	189	83,0
IV q. 2012	14.295	19.000	107	150	0	190	73,0

Source: Authors

- Perform a visual inspection of the residuals obtained from the general regression, (supplemented by the application of the "three-sigma rule").

The first two of the above tools serve as the preliminary tests for the outliers presence in case of the single variables only. However, the third method is much more applicable for detecting so-called multivariate outliers. Given that our aim is to build a multivariate regression model, we will put much more emphasis on the indications obtained from this third approach to outlier detection.

The presence of influential outliers can significantly distort the computations and reduce the credibility of all the statistical tools applied in further phases of the model building process. Therefore, only after identifying any potential outliers and after dealing with them, we will proceed to the next step (that is, to the selection of the final set of the explanatory variables).

3.3.1 The Comparison of Medians and Arithmetic Means of All the Variables

Table 3.3 presents and compares the values of medians and arithmetic means of all the variables.

As we can infer from the above table, in case of all the variables, the distance between the median and the arithmetic mean seems to be rather small. If we assume an arbitrary threshold at 5%, then we can conclude that there are no outlying observations in our dataset. So, our preliminary finding is that the distributions of the individual variables seem to be symmetrical and probably there are not any influential outliers, or if there are, they are offsetting themselves on each side of the distribution.

3.3.2 The Three-Sigma Rule Applied to the Individual Variables

Our next step is to apply the three-sigma rule to each of our seven variables. Table 3.4 presents the results of such analysis.

Table 3.4 shows that all the observations of the individual variables lie within the range set by "three-sigma." This means that the smallest value of each variable exceeds its corresponding lower bound, and the largest value of each variable is lower than its upper bound.

However, the range determined in accordance to the "three-sigma rule" is very conservative, since under the assumption of a normal distribution it covers more than 99% of observations. It means that only some "extreme outliers," that is, only about 0,27% of the observations, those with extremely low or extremely high values, would lie outside this range and many much more common "normal outliers" (suspiciously low or high values, but not "extremely" low or high values) are

Table 3.3 Medians and arithmetic means of the dependent variable as well as all the explanatory variables

Measure	Dependent variable	Candidate explanatory variables					
	$COGS_t$	NS_t	PRM_t	$APOS_t$	$NPOS_t$	NOE_t	SOH_t
	Quarterly costs of goods sold in thousands USD	Quarterly net sales in thousands USD	Wholesale price of basic raw materials in USD/pack.	Total number of all POS at the end of the quarter	Number of the new POS opened in the quarter	Total number of employees at the end of the quarter	Share of hamburgers in the total net sales' revenues (%)
Median	13.343	18.000	109	139	3	170	77,5
Arithmetic mean	12.989	17.385	109	133	3	163	77,2
Median/ mean	2,7%	3,5%	−0,2%	4,6%	0,0%	3,8%	0,4
Mean/ median	−2,7%	−3,4%	0,2%	−4,4%	0,0%	−3,6%	−0,4

Source: Authors

Table 3.4 "Three-sigma rule" applied to the dependent variable as well as all the explanatory variables

Measure	Dependent variable	Candidate explanatory variables					
	$COGS_t$ Quarterly costs of goods sold in thousands USD	NS_t Quarterly net sales in thousands USD	PRM_t Wholesale price of basic raw materials in USD/pack.	$APOS_t$ Total number of all POS at the end of the quarter	$NPOS_t$ Number of the new POS opened in the quarter	NOE_t Total number of employees at the end of the quarter	SOH_t Share of hamburgers in the total net sales' revenues (%)
Arithmetic mean (AM)	12.989	17.385	109	133	3	163	77,2
Std. deviation (SD)	1.969	2.950	4	17	2	23	2,9
AM + 2 × SD	16.927	23.285	116	168	6	210	83,0
AM − 2 × SD	9.050	11.485	101	98	−1	117	71,5
AM + 3 × SD	18.897	26.235	120	185	8	233	85,9
AM − 3 × SD	7.080	8.535	97	80	−3	94	68,6
Max in the sample	15.765	21.900	115	150	6	190	83,0
Min in the sample	8.700	11.200	100	102	0	123	73,0

Source: Authors

Table 3.5 Minimum values and lower bounds based on the two-sigma rule of the dependent variable as well as two explanatory variables

	$COGS_t$	NS_t	PRM_t
Lower bound (arithmetic mean minus two standard deviations)	9.050	11.485	101
Min in the sample	8.700	11.200	100
Min in the sample/lower bound	−3,9%	−2,5%	−1,1%

Source: Authors

still included in the interval determined by the "three-sigma" range. Therefore, it is always justified to check whether all the observations lie within the range set by "two-sigma."

From the above table, we can infer that in the case of $APOS_t$, $NPOS_t$, NOE_t, and SOH_t variables, all the observations lie within the range set by "two-sigma." This means that the smallest values of these variables exceed their corresponding lower bounds and the largest values of these variables are smaller than their respective upper bounds, although some observations are on the verge of breaching the upper bounds. However, in the case of $COGS_t$, NS_t, and PRM_t the smallest values in the samples lie below their lower bounds as determined by "two-sigma." But from the original table with the data (Table 3.2), we can see that all this relates to the first observation from the 1st quarter of 2008. We can also see that the smallest values of $COGS_t$, NS_t, and PRM_t are not much smaller than the lower bounds set by "two-sigma" and all of them lie within the range determined by "three-sigma." This is shown in the table below (Table 3.5).

As we can see, all these three values are only slightly below (less than 5%) their respective lower bounds (as determined by "two-sigma"). So for the time being, we state that there are somewhat "suspicious" and we will be vigilant in our further analysis, but we do not treat these observations as true outliers.

3.3.3 The Analysis of the Fitted Values and the Residuals from the General Regression

To check whether there are any significant multivariate outliers we ran the general regression, in which all the candidate explanatory variables are present. Then we scrutinized the obtained residuals for the presence of any outliers.

After regressing $COGS_t$ against the set of candidate explanatory variables, we obtained the following model (here we provide only the structural parameters):

$$COGS_t = -8.047, 16 + 0, 49\, NS_t + 93, 50\, PRM_t - 53, 95\, APOS_t - 52, 54\, NPOS_t + 50, 00\, NOE_t + 1.852, 00\, SOH_t$$

Then, for the obtained general regression, we computed the fitted values of the dependent variable and we plotted them on the scatter plot, against the actual values of the dependent variable. This is shown in Fig. 3.3.

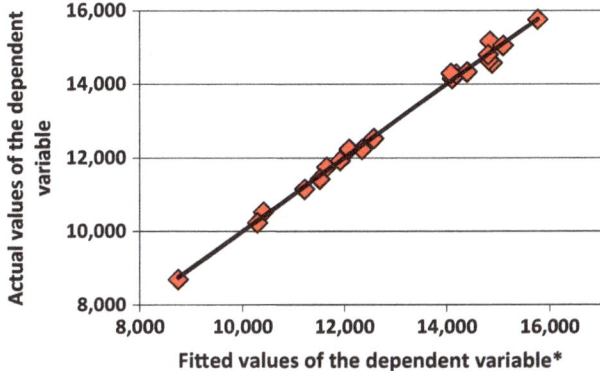

Fig. 3.3 Actual values of the dependent variable against the fitted values of the dependent variable from the general regression.*From the general regression. Source: Authors

It seems that there are not any actual observations lying distinctively far from the line of the fitted values. Chapter 2 shows that such deviations can significantly distort the values of regression coefficients. Also, it seems that there are no observations lying distinctively and influentially far from any cluster of the remaining observations. This is not surprising, because such observations would need to have extremely low or high values of the dependent variable and they probably would have been detected in the previous step by the "three-sigma rule" or the "two-sigma rule." We can also see that the observation indicated in the previous section as "suspicious" by the "two-sigma rule" (from the first quarter of 2008) seems not to influence the whole relationship. We conclude that the relationship shown above, between the actual and fitted values of the dependent variable, seems to be "sound" and not dominated by any single outlying or influential observations.

The residuals obtained from the general regression are presented in Fig. 3.4. Its visual inspection suggests that there are at least two "suspicious" residuals. The first one relates to the 4th quarter of 2010, with a residual larger than 300, much larger than in the case of any other observation. The second one relates to the 2nd quarter of 2011, with a residual smaller than −300. The observation in the 4th quarter of 2012 could perhaps also be considered to be an outlier; however, for further analysis we will treat only the 4th quarter of 2012 and the 2nd quarter of 2011 as potential outliers. In contrast, we can infer from the above figure that the observation from the first quarter of 2008, indicated in the previous section as "suspicious" by the "two-sigma rule," is not producing a significant residual, confirming that it is not an outlier.

One way to deal with the outliers is to add additional dummy variables with value of 1 in case of the outlying observation and 0 otherwise, to the original set of explanatory variables. We visually identified two potential outliers, so we extend our set of candidate explanatory variables by including the following dummy variables:

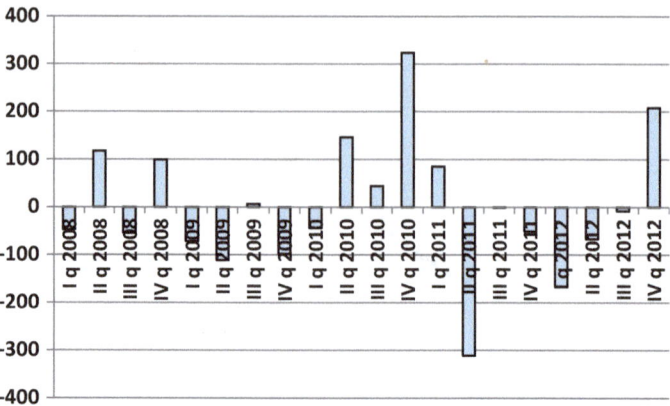

Fig. 3.4 The residuals obtained from the general regression. Source: Authors

• Dummy variable for the 4th quarter of 2010, denoted as IVq2010 and having the
  following values of observations:

$$IVq2010 = \begin{cases} 1 & \text{in the 4th quarter of 2010} \\ & 0 \text{ otherwise} \end{cases}$$

• Dummy variable for the 2nd quarter of 2011, denoted as IIq2011 and having the
  following values of observations:

$$IIq2011 = \begin{cases} 1 & \text{in the 2nd quarter of 2011} \\ & 0 \text{ otherwise} \end{cases}$$

The final inference about whether the observations on the 4th quarter of 2012
and on the 2nd quarter of 2011 constitute the real multivariate outliers will be
obtained in the next step of our model building process, during the selection of the
final set of the explanatory variables. In that next step, we will remove all the
candidate explanatory variables which will turn out not to be statistically significant. Therefore, if any of these two dummy variables, constituting our proxies for
the potential outliers, will "survive" after the variables' selection procedures, this
will mean that this observation should be deemed as a true outlier. However, if these
variables turn out to be statistically insignificant, then these observations were only
"suspects" and they are now "cleared from the accusations" and no longer treated as
outliers.

Before we proceed further, our updated data set looks as follows (Table 3.6).

Table 3.6 The statistical data for all the observations of the dependent variable and all the candidate explanatory variables

| Period | Dependent variable | Candidate explanatory variables | | | | | | | |
| | COGS_t | NS_t | PRM_t | APOS_t | NPOS_t | NOE_t | SOH_t | IVq2010 | IIq2011 |
	Quarterly costs of goods sold in thousands USD	Quarterly net sales in thousands USD	Wholesale price of basic raw materials in USD/pack.	Total number of all POS at the end of the quarter	Number of the new POS opened in the quarter	Total number of employees at the end of the quarter	Share of hamburgers in the total net sales' revenues (%)	Dummy variable for the potential outlier at the 4th quarter of 2010	Dummy variable for the potential outlier at the 2nd quarter of 2011
I q. 2008	8.700	11.200	100	102	2	123	74,5	0	0
II q. 2008	10.525	13.400	105	104	2	126	78,0	0	0
III q. 2008	10.235	12.700	107	107	3	130	81,0	0	0
IV q. 2008	11.750	15.000	110	110	3	134	76,0	0	0
I q. 2009	11.145	14.500	109	115	5	140	75,0	0	0
II q. 2009	12.235	16.900	107	119	4	145	77,5	0	0
III q. 2009	11.925	16.400	105	124	5	151	79,0	0	0
IV q. 2009	11.420	16.000	104	130	6	158	74,5	0	0
I q. 2010	12.530	17.500	106	135	5	164	77,0	0	0
II q. 2010	12.240	15.800	108	138	3	168	78,5	0	0
III q. 2010	14.155	19.000	111	140	2	171	82,0	0	0
IV q. 2010	15.165	20.300	113	143	3	175	77,5	1	0
I q. 2011	14.275	18.500	115	145	2	178	75,0	0	0
II q. 2011	14.575	19.600	115	146	1	180	78,0	0	1
III q. 2011	15.765	21.900	113	149	3	184	78,0	0	0
IV q. 2011	15.050	21.000	110	150	1	186	73,0	0	0
I q. 2012	14.655	20.000	111	150	0	187	74,0	0	0
II q. 2012	14.330	19.000	110	150	0	188	80,0	0	0
III q. 2012	14.800	20.000	108	150	0	189	83,0	0	0
IV q. 2012	14.295	19.000	107	150	0	190	73,0	0	0

Source: Authors

3.4   Selection of Explanatory Variables (From the Set of Candidates)

After we completed the process of creating the set of candidate explanatory variables (including dummy variables representing the potential outliers), we can proceed to selecting the final set of explanatory variables. The aim of this selection is to leave in the final set only those variables which:

- Have statistically significant relationships with the dependent variable.
- Are not conveying (copying) the same information about the dependent variable.

The logic behind the first of the above points is straightforward: it makes no sense to hold in the model those explanatory variables which are not statistically related to the dependent variable. If they do not show any statistically significant relationships with the dependent variable, they cannot tell us anything interesting about this variable. They should be removed from further analysis.

The second point is more subtle. It means that if two or more candidate variables carry the same information about the dependent variable, then keeping all of these explanatory variables in the model is unnecessary. If several explanatory variables are related to the dependent variable in a similar way, which usually also means that these explanatory variables are strongly correlated with each other, then it is sufficient to leave in the model only one of them as the proxy for the information carried by all of them. Holding strongly inter-correlated explanatory variables in the model is also problematic from the point of view of estimation procedures, because the resulting multicollinearity significantly distorts the standard errors of model coefficients, and in extreme cases can even break down the estimation procedure. Fortunately, both procedures of variable selection are usually effective in dealing with the above problems.

For our model, we will apply two alternative procedures for the selection of explanatory variables:

- "General to specific modeling," which begins with the general model. The whole set of candidate explanatory variables is initially included and then reduced step by step by removing the insignificant variables, one by one.
- "Stepwise regression," which runs exactly opposite to the "general to specific modeling." This method begins with the set of most reduced, univariate regressions and then adds the most significant of the remaining variables one by one.

3.4.1   The Procedure of "General to Specific Modeling"

The procedure of "general to specific modeling" runs as is shown in Table 3.7 (Davidson et al. 1978; Charemza and Deadman 1992).

Table 3.7 The procedure of "general to specific modeling"

Step 1	Estimate the parameters of the general model
Step 2	Check whether each explanatory variable in the general model is statistically significant (e.g., by means of t-Statistics)
Step 3	If all the explanatory variables are statistically significant, leave the model in its current form (that is, with all the initial variables)
Step 4	Remove from the model the variable which is the least statistically significant (e.g., has the lowest absolute value of t-Statistic)
Step 5	Reestimate the model without the removed explanatory variable
Step 6	Return back to Steps 2–5
Step 7	Continue the procedure until You end up with the model in which all the final explanatory variables are statistically significant

Source: Authors

For simplicity, we will base our evaluation of the variables significance on the simplest and most popular criteria, the t-Statistic. However, it should be kept in mind that this criteria can be misleading in the presence of one or more of the following: autocorrelation of residuals, heteroscedasticity of residuals, or multicollinearity of explanatory variables. In such cases, some more advanced statistical tools should be applied (which beyond the scope of this book).

At the beginning, there are 8 candidate explanatory variables and only 20 observations. This implies relatively few degrees of freedom and hence an increased risk of an upward bias in the absolute values of t-Statistics. We assume a 1% significance level for the critical values of t-Statistics, instead of the more common approach based on 5% significance level.

For our general model, we obtained the following results (Table 3.8).

Given the results shown in the above table, we should remove $APOS_t$ from our set of explanatory variables on the ground that the absolute value of its t-Statistic is below the critical value and is the lowest among all the explanatory variables. After removal of $APOS_t$, we reestimate the parameters of the reduced model and repeat the procedure of the statistical significance analysis. The results of this re-estimation are presented below (Table 3.9).

Again, on the ground of the above results, we should remove SOH_t from the set of explanatory variables, because it has the absolute value of t-Statistic below the critical value and the lowest among all the explanatory variables. We again reestimate the parameters of the reduced model after removal of SOH_t and repeat the procedure of the statistical significance analysis. The results of this second re-estimation are presented below (Table 3.10).

After removing $IVq2010$ from the set of explanatory variables, we again reestimate the parameters of this further reduced model. The results of this third re-estimation are presented below (Table 3.11).

After removing NOE_t from the set of explanatory variables, we again reestimate the parameters of this further reduced model. The results of this fourth re-estimation are presented below (Table 3.12).

Table 3.8 Regression results for the general model (i.e., model with the full set of all the candidate explanatory variables)

Variable	Coefficient	t-Statistic[a]
Intercept	−8.795,24	−6,620
NS$_t$	0,47	16,317
PRM$_t$	101,13	8,682
APOS$_t$	−56,26	**−1,762**
NPOS$_t$	−59,79	−2,375
NOE$_t$	53,74	2,154
SOH$_t$	1.931,77	2,058
IVq2010	345,30	2,620
IIq2011	−348,10	−2,573

R-squared: 1,00
Critical value of t-Statistic[b]: 3,106

Conclusion: remove APOS$_t$

Source: Authors
[a]The smallest absolute value is noted
[b]For 20 observations, 8 explanatory variables, and 1% significance level

Table 3.9 Regression results for the reestimated model (i.e., after removing APOS$_t$ from the set of explanatory variables)

Variable	Coefficient	t-Statistic[a]
Intercept	−8.405,44	−5,918
NS$_t$	0,47	15,345
PRM$_t$	94,76	7,893
NPOS$_t$	−90,94	−4,681
NOE$_t$	10,22	2,737
SOH$_t$	1.822,81	**1,795**
IVq2010	303,32	2,158
IIq2011	−386,18	−2,668

R-squared: 1,00
Critical value of t-Statistic[b]: 3,055

Conclusion: remove SOH$_t$

Source: Authors
[a]The smallest absolute value is noted
[b]For 20 observations, 7 explanatory variables and 1% significance level

Again, we need to remove the statistically insignificant variable with the lowest absolute value of t-Statistic. This time this is IIq2011. After its removal we need to once again reestimate the parameters of the reduced model. The results of this fifth re-estimation are presented below (Table 3.13).

We can see that all three remaining explanatory variables are statistically significant at 1% significance level. This allows us to abort the procedure of variable selection. We began with a general model which included all eight candidate explanatory variables, and we ended up with the specific model including only the statistically significant explanatory variables: NS$_t$, PRM$_t$, and NPOS$_t$.

Table 3.10 Regression results for the reestimated model (i.e., after removing SOH_t from the set of explanatory variables)

Variable	Coefficient	t-Statistic[a]
Intercept	−7.060,61	−5,408
NS_t	0,48	14,305
PRM_t	95,79	7,382
$NPOS_t$	−93,76	−4,475
NOE_t	9,61	2,389
IVq2010	303,09	**1,993**
IIq2011	−379,62	−2,424
R-squared: 1,00		
Critical value of t-Statistic[b]: 3,012		
Conclusion: remove IVq2010		

Source: Authors
[a]The smallest absolute value is noted
[b]For 20 observations, 6 explanatory variables and 1% significance level

Table 3.11 Regression results for the reestimated model (i.e., after removing IVq2010 from the set of explanatory variables)

Variable	Coefficient	t-Statistic[a]
Intercept	−7.529,02	−5,325
NS_t	0,49	13,583
PRM_t	100,19	7,116
$NPOS_t$	−90,37	−3,931
NOE_t	8,26	**1,892**
IIq2011	−425,20	−2,493
R-squared: 1,00		
Critical value of t-Statistic[b]: 2,977		
Conclusion: remove NOE_t		

Source: Authors
[a]The smallest absolute value is noted
[b]For 20 observations, 5 explanatory variables, and 1% significance level

Table 3.12 Regression results for the reestimated model (i.e., after removing NOE_t from the set of explanatory variables)

Variable	Coefficient	t-Statistic[a]
Intercept	−6.548,81	−4,598
NS_t	0,55	29,510
PRM_t	94,43	6,346
$NPOS_t$	−109,93	−4,944
IIq2011	−413,06	**−2,239**
R-squared: 0,99		
Critical value of t-Statistic[b]: 2,947		
Conclusion: remove IIq2011		

Source: Authors
[a]The smallest absolute value is noted
[b]For 20 observations, 4 explanatory variables, and 1% significance level

Table 3.13 Regression results for the reestimated model (i.e., after removing IIq2011 from the set of explanatory variables)

Variable	Coefficient	t-Statistic[a]
Intercept	−5.350,86	−3,625
NS$_t$	0,56	27,146
PRM$_t$	81,92	5,312
NPOS$_t$	−104,01	**−4,213**

R-squared: 0,99
Critical value of t-Statistic[b]: 2,921

Conclusion: all the explanatory variables are statistically significant at 1% significance level

Source: Authors
[a]The smallest absolute value is noted
[b]For 20 observations, 3 explanatory variables and 1% significance level

Finally, our linear regression model, obtained from the "general to specific modeling" procedure, looks as follows:

$$COGS_t = -5.350,86 + 0,56\ NS_t + 81,92\ PRM_t - 104,01\ NPOS_t$$

We will conduct the interpretation of the obtained model in the next section.

3.4.2   The Procedure of "Stepwise Regression"

The procedure of "stepwise regression" runs as is shown in Table 3.14 (Nilsson and Nilsson 1994).

Although there are many tools for evaluating the goodness of fit of the model, we will for simplicity base our evaluation on probably the most popular criteria, which is R-squared.

Given that at the beginning we have eight candidate explanatory variables, we begin with eight different univariate regressions. Each regression at this stage has the intercept and one slope parameter; however, we will present only the R-squared statistics for all the individual regressions and t-Statistics for the slope parameters.

For this set of eight alternative regression, we obtained the following results (Table 3.15).

As we can see from the above table, the variable with the strongest statistical relationship with the dependent variable is NS$_t$. It can be inferred both from the highest R-squared among all the eight univariate regressions (0,971) as well as from the largest absolute value of t-Statistic (24,501). After picking NS$_t$ to the following step, we need to estimate seven regressions. In each of these regressions, there will be two explanatory variables: NS$_t$ and one of the other candidate explanatory variables. The results of such two-variable regressions are presented below (Table 3.16).

Table 3.14 The procedure of "stepwise regression"

Step 1	If You have i candidate explanatory variables, estimate i independent simple (univariate) regressions
Step 2	Pick the regression with the largest value of a selected goodness-of-fit statistic (e.g., R-squared)
Step 3	Estimate $i - 1$ two-variable regressions by adding the $i - 1$ individual remaining candidate explanatory variables to the univariate regression chosen in Step 2
Step 4	From the $i - 1$ two-variable regressions built in Step 3. Pick the one with the highest value of a selected goodness-of-fit statistic
Step 5	Estimate $i - 2$ three-variable regressions by adding the $i - 2$ individual remaining candidate explanatory variables to the two-variable regression chosen in Step 4
Step 6	Continue the procedure until the moment when no more added candidate explanatory variables are statistically significant (or until the model gets too large)

Source: Authors

Table 3.15 Regression results for the univariate models

Candidate variables in the regression	R-squared[a]	t-Statistics[b]
NS_t	**0,971**	24,501
PRM_t	0,600	5,200
$APOS_t$	0,862	10,598
$NPOS_t$	0,235	−2,353
NOE_t	0,859	10,466
SOH_t	0,005	**0,300**
IVq2010	0,068	**1,143**
IIq2011	0,036	**0,819**
Critical value of t-Statistic[c]: 2,878		
Conclusion: pick NS_t		

Source: Authors
[a]The highest value is noted
[b]The statistically insignificant values are noted
[c]For 20 observations, 1 explanatory variable, and 1% significance level

On the ground of the above data, we're picking PRM_t before we proceed to the next step. The two-variable regression with NS_t and PRM_t as explanatory variables has the highest R-squared among all the seven two-variable regressions. This is also the only one regression in which both slope parameters are statistically significant. After picking PRM_t for the following step, we need to estimate six regressions. In each of these regressions, there will be three explanatory variables: NS_t and PRM_t (picked in the previous steps) and one of the other candidate explanatory variables. The results of such three-variable regressions are presented below (Table 3.17).

Now we're picking $NPOS_t$, because the three-variable regression with NS_t, PRM_t, and $NPOS_t$ as explanatory variables has the largest value of R-squared among all the six three-variable regressions. After picking $NPOS_t$, we estimate five regressions. In each of these regressions, there will be four explanatory

Table 3.16 Regression results for the two-variable models

Candidate variables in the regression	R-squared[a]	t-Statistics[b]
NS_t, PRM_t	**0,985**	20,776; 3,960
NS_t, $APOS_t$	0,973	8,277; **1,012**
NS_t, $NPOS_t$	0,980	25,265; **−2,819**
NS_t, NOE_t	0,973	8,507; **1,199**
NS_t, SOH_t	0,972	24,331; **0,905**
NS_t, IVq2010	0,972	23,389; **0,783**
NS_t, IIq2011	0,971	23,468; **0,381**
Critical value of t-Statistic[c]: 2,898		
Conclusion: pick PRM_t		

Source: Authors
[a]The highest value is noted
[b]The statistically insignificant values are noted
[c]For 20 observations, 2 explanatory variables, and 1% significance level

Table 3.17 Regression results for the three-variable models

Candidate variables in the regression	R-squared[a]	t-Statistics[b]
NS_t, PRM_t, $APOS_t$	0,988	8,901; 4,549; **2,058**
NS_t, PRM_t, $NPOS_t$	**0,993**	27,146; 5,312; −4,213
NS_t, PRM_t, NOE_t	0,990	9,226; 5,016; **2,688**
NS_t, PRM_t, SOH_t	0,986	20,656; 3,842; **0,880**
NS_t, PRM_t, IVq2010	0,985	20,218; 3,755; **0,470**
NS_t, PRM_t, IIq2011	0,986	20,467; 4,079; **−1,059**
Critical value of t-Statistic[c]: 2,921		
Conclusion: pick $NPOS_t$		

Source: Authors
[a]The highest value is noted
[b]The statistically insignificant values are noted
[c]For 20 observations, 3 explanatory variables, and 1% significance level

variables: NS_t, PRM_t, and $NPOS_t$ (picked in the previous steps) and one of the other candidate explanatory variables. The results of such four-variable regressions are presented below (Table 3.18).

From the above table, we can see that the regression with the highest goodness of fit is the one with NS_t, PRM_t, $NPOS_t$, and IVq2011 as the explanatory variables. However, we can also note that in all five regressions the last slope parameter, related to the variable added in this step, is statistically insignificant. It means that we can abort the procedure of "stepwise regression" and we can conclude that our final set of explanatory variables is the one which was determined in the previous step (that is with NS_t, PRM_t, and $NPOS_t$ as the only explanatory variables).

Finally, we may see that the results of this "stepwise regression" are identical to the results of the "general to specific modeling" procedure. This means that our

Table 3.18 Regression results for the four-variable models

Candidate variables in the regression	R-squared[a]	t-Statistics[b]
$NS_t, PRM_t, NPOS_t, APOS_t$	0,994	12,248; 5,598; −3,621; **1,363**
$NS_t, PRM_t, NPOS_t, NOE_t$	0,994	12,030; 5,744; −3,207; **1,550**
$NS_t, PRM_t, NPOS_t, SOH_t$	0,993	27,257; 5,223; −4,173; **1,053**
$NS_t, PRM_t, NPOS_t, IVq2010$	0,994	27,977; 5,218; −4,615; **1,564**
$NS_t, PRM_t, NPOS_t, IIq2011$	**0,995**	29,510; 6,346; −4,944; **−2,239**
Critical value of t-Statistic[c]: 2,947		
Conclusion: abort the procedure of picking variables		

Source: Authors
[a]The highest value is noted
[b]The statistically insignificant values are noted
[c]For 20 observations, 4 explanatory variables, and 1% significance level

linear regression model, obtained from both procedures of variables' selection, looks as follows:

$$COGS_t = -5.350,86 + 0,56\,NS_t + 81,92\ PRM_t - 104,01\,NPOS_t$$

We will conduct the interpretation of the obtained model in the next section.

3.4.3   Which Procedure to Apply?

Both procedures produced identical final results. One could wonder, therefore, why we illustrated the two procedures and not just one of them. Both procedures differ significantly in terms of their time consumption and applicability in different situations.

The "general to specific modeling" is generally less time-consuming than the "stepwise regression." If You have i candidate explanatory variables, than You will need to reestimate Your regression no more than $i - 1$ times if modeling from general model to the specific one. In contrast, under the "stepwise regression" You will need to estimate i univariate regressions in just the first step (and $i - 1$ in the second step, $i - 2$ it the third step, etc.). So, if You have, say, 20 candidate explanatory variables, You will have to estimate no more than 19 regressions (from general to specific) and if You applied the "stepwise regression" in the same circumstance, then You would perhaps need to estimate as many as 210 regressions (i.e., 20 in the first step, 19 in the second step, 18 in the first step, etc.). So, from the point of view of time consumption, the "general to specific modeling" procedure is usually superior to the "stepwise regression."

However, in many practical applications the "general to specific modeling" has severe limitations, related to the required (for maintaining the statistical credibility) degrees of freedom. The number of degrees of freedom is the difference between the number of observations and the number of estimated parameters. So, for example, if You have 20 observations and, say, 15 candidate explanatory variables, then the number of degrees of freedom in the first step is very small. This results in obtaining the unreliable (usually highly exaggerated) t-Statistics for the individual slope parameters as well as heavily overstated R-squared. This, in turn, may result in getting a spurious specific model, involving explanatory variables with allegedly significant statistical relationships with the dependent variable, while many of them can be statistically insignificant in reality. Also, by including all the candidate explanatory variables into the general model, one risks receiving unreliable results caused by the multicollinearities among the individual explanatory variables, which also significantly distort the regression statistics. So, from the point of view of statistical reliability, the "stepwise regression," although more time-consuming, is usually more credible than the "general to specific modeling."

Given the above discussion, we recommend application of the "general to specific modeling" when all of the following conditions are satisfied:

- The number of degrees of freedom is at least double-digit.
- The total number of candidate explanatory variables is not too large (as a rule of thumb, let's assume, no more than 15 variables).
- The analyst's knowledge and experience suggests that the probability of having strongly intercorrelated candidate explanatory variables is low.

If any of the above conditions is not met, then we recommend applying the "stepwise regression."

3.5 Interpretation of the Obtained Regression Structural Parameters

After constructing the econometric model, the next step is to verify its correctness and reliability. Then we can proceed to the interpretation of the obtained model and to its practical application.

The full-blown interpretation and application of the model should be always preceded by its verification (which we discuss with details in the following chapter). Therefore, at this stage, we will provide only the tentative interpretation of the structural parameters of our model.

Let's remind that our model (at this stage) looks as follows:

$$COGS_t = -5.350, 86 + 0, 56 \, NS_t + 81, 92 \, PRM_t - 104, 01 \, NPOS_t$$

where:

COGS$_t$—quarterly cost of goods sold (in thousands of USD),
NS$_t$—quarterly net sales (in thousands of USD),
PRM$_t$—quarterly wholesale price of raw materials (beef) per package (in USD),
NPOS$_t$—the number of new points of sale (POS) opened in a given quarter.

In this case, the negative value of intercept in our model cannot be interpreted sensibly. Normally, the value of the intercept is deemed to reflect the expected value of the dependent variable in the case when the values of all the explanatory variables are set at zero. Very often this value seems to make sense and is being interpreted somehow, while it should not be. In our case, the obtained negative value (equaling −5.350,86) would literally mean that if:

- The company in a given quarter does not generate any net sales revenues.
- The quarterly average price of the raw materials (meat) in the same quarter equals zero.
- The company in the same quarter does not open any new points of sale, then the quarterly costs of goods sold is −5.350,86 thousands of USD.

It would literally mean that the best way to reorganize this business would be to stop any sales, to cease opening any new points of sales and to ensure that the meat being purchased is obtained for free! Under such a scenario, the company would not have any net sales, but would have the *negative* costs of goods sold (equaling −5.350,86 thousands of USD), which would imply a *positive* gross margin on sales (the difference between net sales and costs of goods sold) equaling 5.350,86 thousands of USD (net sales less costs of goods sold = 0 + 5.350,86).

Such an interpretation does not make any *economical* sense. That is why it is always very important not to make any automatic (mechanical) interpretation and application of any statistical and econometric models. Instead, the statistical analysis must always be accompanied by a rigorous consideration of any relevant economical (and social, legal, technological, etc.) factors.

In the case of models similar to ours, in which the dependent variable is cost of goods sold, the risk of deriving erroneous inference is particularly high, because models like that are often recommended by some books on management accounting (and cost accounting), as the tools for splitting the company's total operating costs into fixed costs and variable costs. In such models, the intercept is deemed to reflect the average fixed operating costs, and the slope coefficient (e.g., at net sales or production in units) is interpreted as the average increase (in USD) of the total operating costs brought about by the increase of the production or sales volume by one unit. Models like these can be reliably applied in making the cost' breakdowns only when we know from our experience or from other sources that the obtained intercept is not significantly overstated or understated.

In turn, the significant overstatement or understatement of the obtained intercept can be caused by at least two factors (discussed and illustrated with more details in the following chapters):

- The structural changes in the underlying relationships between the dependent variable and the explanatory variables, occurring within the sample on the basis of which the model was estimated.
- The nonlinearities of the relationships between the dependent variable and the explanatory variables, which can occur both within the sample and out-of-sample.

Additionally, the literal interpretation of the intercept, as the expected value of the dependent variable when the values of all explanatory variables are equal to zero, implies the assumption that such a scenario of zero values of all the explanatory variables at the same time can happen. Although we can imagine that our company does not open any new points of sales and even temporarily does not generate any sales revenues, one would be rather mad to seriously consider the possibility of getting such a bargain price of raw meat, as equal to zero USD per pack. Well, it's not a good thing to simulate the real-life business scenarios by taking fiction-based assumptions.

To sum up, it is obvious that in our model the obtained negative value of the intercept should not be economically interpreted. However, this coefficient is statistically significant (having the absolute value of t-Statistic of 3,625 when the critical value, at 1% level of statistical significance, is 2,921), so from the statistical point of view it is justified to leave this parameter in the model.

Let's now turn to the interpretation of the obtained slope coefficients. The tentative interpretation suggests that:

- The quarterly costs of goods sold tend to increase/decrease by USD 0,56, on average, with every one dollar of increased/decreased quarterly net sales.
- The quarterly costs of goods sold tend to increase/decrease by 81,92 thousands of USD, on average, with every one dollar of increase/decrease of the wholesale price of basic raw materials (per pack).
- The quarterly costs of goods sold tend to decrease/increase by 104,01 thousands of USD, on average, with every new point of sale open in a given quarter.

The direction of the relationships between the dependent variable and the first two of the explanatory variables (i.e., net sales and the price of raw meat) seems to be fully intuitive and according to expectations. The higher/lower the sales volume, as measured by the net sales revenues, the higher/lower the costs of goods sold, which must be incurred. The higher/lower the price of raw materials, the higher/lower the costs of goods sold (because if we have to pay more for the supplies purchased, then the total costs increase, even if the physical volume and structure of sales do not change).

However, the obtained negative direction of the averaged statistical relationship between the costs of goods sold and the number of new points of sales opened in the quarter is much more surprising and challenging. On its face a relationship like this should be positive, meaning that the more new points of sale, the higher the costs. This is so because of the impact of at least some of the following factors:

- Just after opening a new point of sale some of the costs, especially those of fixed nature, are already being incurred in full (e.g., electric energy used by the refrigerators or fixed part of the sales staff remuneration), while the sales can still be at the moderate start-up level, because the point of sale can be not yet recognized by many consumers—this suggests the temporary deterioration of the costs–sales relationship just after the new POS openings.

- Even if the physical volume of sales is relatively high just from the beginning, perhaps because of a good location of the new POS and its resulting good visibility, it is often supported by some promotional incentives used by the company at this start-up phase, such as discount prices for the first few days, happy hours, or bonuses for repeated sales. Often such practices are intensified in the early stages of the sales point, which again suggests the temporary distortions of the company's long-term costs–sales relationships,
- Often the effectiveness of the new point of sales in many areas, such as inventory management, time of consumer service, and product quality, is relatively poor as compared to more "experienced" sales points, which is due to the effect of so-called "learning curve"—this again suggests relatively high unit costs (and as a result also total costs) of goods sold in the early stage of the new sales points.

The above factors are only examples of the forces which can cause the subdued gross margins in the periods around the new sales' points openings. All suggest rather positive relationships between the number of new points of sales and the costs of goods sold in a period. Therefore, the obtained negative relationship in our model is quite puzzling and perhaps requires deeper analysis of the marketing and operating factors specific for the company under investigation.

References

Charemza WW, Deadman DF (1992) New directions in econometric practice. General to specific modelling, cointegration and vector autoregression. Edward Elgar, Aldershot
Davidson JH, Hendry DH, Srba F, Yeo S (1978) Econometric modelling of the aggregate time-series relationship between consumers' expenditure and income in the United Kingdom. Econ J 88:661–692
Nilsson C, Nilsson J (1994) A time series approach to selecting inflation indicators, Sveriges Riksbank Arbetsrapport. Sveriges Riksbank, Stockholm

Chapter 4
Verification of the Multiple Regression Model

4.1 Introduction

In the last section of the previous chapter, we provided the tentative interpretation of structural parameters of our model. However, before the model is fully interpreted and applied in practice for forecasting or making simulations, it must be verified for statistical correctness.

In this chapter, we will discuss a comprehensive procedure for model verification, composed of the following tests:

- Tests for goodness of fit to the actual data (based on the coefficient of determination, discussed in Chap. 1).
- F-test for the general statistical significance of the whole model and t-Student tests for the statistical significance of the individual parameters.
- Tests for normality of distribution of residuals (Hellwig test and Jarque–Bera test).
- Tests for autocorrelation of residuals (F-test and Box–Pearce test).
- Tests for heteroscedasticity of residuals (*ARCH-LM* test and Breusch–Pagan test).
- Tests for symmetry of residuals (symmetry test and t-Student test of symmetry).
- Tests for randomness of residuals (maximum series length test and number of series test).
- Test for the general specification of the model (Ramsey's *RESET* test).
- Test for multicollinearity of explanatory variables (variance inflation factor test).

The majority of tests discussed in this chapter are based on residuals from the obtained tested model. In the case of any single observation included in the sample on which the model was estimated, the residual is the difference between actual value of the observation and its fitted value obtained from the estimated model.

A good econometric model not only should have the required goodness of fit, and should be statistically significant, but it also should generate residuals which satisfy some conditions. Generally speaking, the obtained residuals should be normally distributed and as chaotic or noisy as possible, so that we cannot discern

© Springer International Publishing AG 2018
J. Welc, P.J.R. Esquerdo, *Applied Regression Analysis for Business*,
https://doi.org/10.1007/978-3-319-71156-0_4

any regular patterns or "strange" behavior within the sample of residuals. Particularly, a "good" model should have residuals which satisfy the following conditions:

- They are normally distributed and particularly are not having neither skewed distribution nor "fat tails."
- They are not autocorrelated (the residuals do not show any statistical relationships with each other).
- They have a stable variance within the whole sample (they are homoscedastic instead of heteroscedastic).
- They are symmetrically distributed around zero (the number of positive residuals does not differ significantly from the number of negative residuals).
- They are randomly distributed (the sign of any given residual is not related with the signs of adjacent residuals).

Also, the estimated model should have the correct functional form (i.e., linear, exponential, power, etc.), consistent with the actual functional form of the modeled relationships and should not include explanatory variables which are strongly correlated with each other.

In practice, it is often very difficult to construct a model which passes all the tests for its correctness. It is very often the case that the residuals do not have normal distribution or are significantly autocorrelated or are highly heteroscedastic. What's worse, often it is not possible to correct the estimated model (because of lack of necessary data or lack of any knowledge suggesting the way of correction). However, even when the analyst is not able to correct the model, the detailed knowledge of the model flaws is very important for model application and interpretation.

In the following sections, we will discuss the statistical tests useful in verifying the econometric models. The tools presented are not the only ones which can serve the process of model verification. The interested readers can find many other tests in other books. We have decided to include in this book only the tools which can be easily applied with the use of common spreadsheets (e.g., MS Excel or Open Office) and do not require any specialized software.

For most of the potential problems tested, we provide two alternative tests. In most of the cases they should be treated as complementary, which means that it is recommended practice to apply both tools simultaneously, wherever possible. If both tests suggest conflicting conclusions, then we recommend to follow the "negative" conclusion. For example, if testing for heteroscedasticity by means of *ARCH-LM* test and Breusch–Pagan test, the following scenarios are possible:

- Both *ARCH-LM* test and Breusch–Pagan test points to lack of heteroscedasticity (then we conclude that there is not any significant heteroscedasticity of residuals).
- Both *ARCH-LM* test and Breusch–Pagan test points to presence of heteroscedasticity (then we conclude that there is significant heteroscedasticity of residuals).

- *ARCH-LM* test points to lack of heteroscedasticity while Breusch–Pagan test points to presence of heteroscedasticity (then it is safer to assume that there is some heteroscedasticity of residuals).
- *ARCH-LM* test points to presence of heteroscedasticity while Breusch–Pagan test points to lack of heteroscedasticity (then it is safer to assume that there is some heteroscedasticity of residuals).

In Sect. 4.10 of this chapter, we provide suggestions on what to do if the process of model verification points to some problems with the correctness of the tested model. In final section, we summarize the verification of our estimated linear model of $COGS_t$ (quarterly costs of goods sold in thousands of USD).

4.2 Testing General Statistical Significance of the Whole Model: F-test

The purpose of this test is to verify the overall statistical significance of the relationships between the whole set of the explanatory variables and the dependent variable. The most popular test for the general significance of the whole regression is the F-test (or Fisher–Snedecor test) (Harvey 1990; Greene 1997; Verbeek 2000).

The F-test serves for verification of only the general significance of the whole set of explanatory variables, and not as a check for the significance of the individual explanatory variables. This means that when the null hypothesis H_0 is rejected, we infer that at least one (perhaps only one) explanatory variable is statistically significant. It can, therefore, happen that in a model with ten explanatory variables only one is significant and the F-test indicates for the statistical significance of the whole regression, even though all the remaining nine explanatory variables are insignificant. In multiple regression, F-test should therefore be always accompanied with the tests for the statistical significance of the individual explanatory variables (Table 4.1).

Our model was estimated on the basis of $n = 20$ observations; there are $k = 3$ explanatory variables and the coefficient of determination, $R^2 = 0,993$. Putting these numbers into the formula for the empirical statistic, F produces the following results:

$$F = \frac{R^2}{1 - R^2} \frac{n - (k + 1)}{k} = \frac{0,993}{1 - 0,993} \frac{20 - (3 + 1)}{3} = 756,57.$$

We put the value of R-squared into the above formula as rounded to three decimal places, while in most spreadsheets this number is taken at much more exact value, thus your result may differ somewhat from ours.

Now we need to derive the critical value of F^*. For $n = 20$ and $k = 3$, we obtain:

$$m_1 = k = 3,$$

Table 4.1 The procedure of F-test for the general statistical significance of the whole set of explanatory variables

Step 1	Formulate the hypotheses: $H_0 : b_1 = b_2 = \ldots = b_k = 0$ (all slope parameters $= 0$) $H_1 : b_1 \neq 0 \vee b_2 \neq 0 \vee \ldots \vee b_k \neq 0$ (at least one slope parameter $\neq 0$) where: $b_1, b_2, \ldots, b_k$—slope parameters in the tested regression, k—number of explanatory variables in the regression.
Step 2	Compute the empirical statistic F: $F = \frac{R^2}{1-R^2} \frac{n-(k+1)}{k}$, where: R^2—coefficient of determination (R-squared) of the tested regression, n—number of observations on which the tested model was estimated, k—number of explanatory variables in the regression.
Step 3	Derive the critical (theoretical) value F^* from the Fisher–Snedecor table, for $m_1 = k$, $m_2 = n - (k+1)$ and for the assumed significance level α.
Step 4	Compare empirical F with critical F^*. If $F \leq F^*$, then H_0 is not rejected, which means that all the slope parameters in the tested model are not statistically different from zero. If $F > F^*$, then H_0 is rejected (and H_1 is true), which means that at least one slope parameter in the tested model is statistically different from zero.

Source: Authors

$$m_2 = n - (k + 1) = 20 - (3 + 1) = 16.$$

For the level of statistical significance set at $\alpha = 0,05$, and computed values of m_1 and m_2, we can derive the critical value of F^* from the Fisher–Snedecor table or from the appropriate Excel function, which equals 3,24.

Finally, we compare the value of empirical statistic F with its critical value F^*. In our case $F > F^*$, which means that H_0 is rejected, and we conclude that at least one slope parameter in our model is statistically different than zero.

Often the following step after testing the general significance of the model is to test the statistical significance of its individual parameters. However, given the procedures for variables' selection, which we have applied (i.e., "general to specific modeling" and "stepwise regression"), the statistical significance of the individual explanatory variables has already been ensured because both procedures of variables' selection were based on the t-Statistics for the individual explanatory variables. Therefore, we do not discuss here the procedures for checking the significance of individual parameters.

4.3   Testing the Normality of Regression Residuals' Distribution

4.3.1   Nature and Relevance of Residuals' Distribution

For the econometric model to be correct, it is important to obtain residuals which are normally distributed. Lack of normality of residuals' distribution can distort the statistics for the model significance. Moreover, if the model is constructed for predictive purposes, the normal distribution of residuals narrows the expected range of forecast errors, increasing the forecasts' accuracy and reduces the risk of obtaining biased predictions.

We will present two types of deviations of empirical distributions from normality:

- Deviation from normality caused by asymmetry (skewness) of distribution
- Deviation from normality caused by so-called fat tails of distribution.

4.3.2   Illustration of Non-normality of Distribution Caused by Asymmetry (Skewness)

Figure 4.1 presents an example of an asymmetric (skewed) distribution of residuals. The normal distribution is shown in blue line and the actual distribution is shown in red line. The vertical axis informs about the frequency of distribution.

Under the normal distribution of residuals, the following conditions are satisfied:

- The arithmetic average of all residuals equals zero.
- The distribution is symmetrical.
- The largest frequency of residuals is clustered around the arithmetic average.

In contrast, the asymmetric (skewed) actual distribution shown in Fig. 4.1 has the following features:

- The arithmetic average of all residuals equals zero (similarly as in case of normal distribution).
- The distribution is skewed to the left, which in this example means that there is some "shortage" of residuals of values between -150 and -80, there is some "excess" of residuals of values between -80 and 0, and there is "shortage" of residuals of values between 0 and 80.
- The skewness to the left also means that there are more negative residuals than positive residuals (here this is 53,4% vs. 46,6%).
- The largest frequency of residuals is clustered to the left from arithmetic average and hovers around -40.

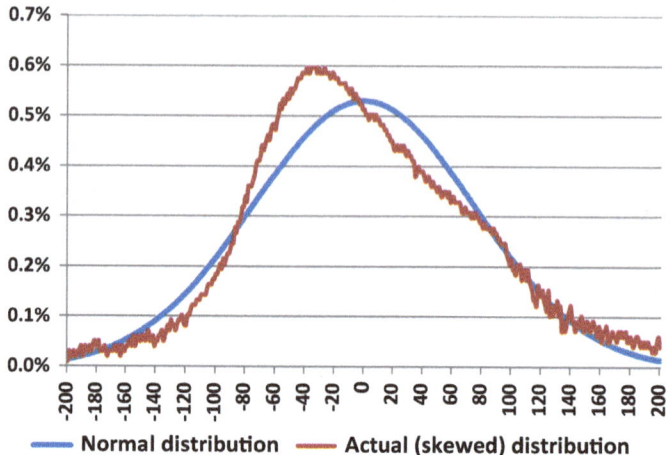

Fig. 4.1 Example of asymmetric (skewed) distribution of residuals. Source: Authors

Such asymmetrical deviation of the actual distribution of residuals from normality results not only in distorted values of some regression statistics, but also in biased estimates of the dependent variable.

4.3.3   Illustration of Non-normality of Distribution Caused by "Fat Tails"

Figure 4.2 presents another example of the non-normal distribution of residuals. Here again the normal distribution is shown in blue and the actual distribution is shown in red. The vertical axis informs about the frequency of distribution.

Opposite to the previous example, the actual distribution here shares some common features with the normal distribution, that is:

- The arithmetic average of all residuals equals zero.
- The distribution is symmetrical.
- The largest frequency of residuals is clustered around the arithmetic average.

However, these characteristics by themselves do not mean that such distribution is normal. Although it is not skewed, is still has some "shortages" of residuals as compared to normal distribution and it has some "excesses" of residuals as compared to normal distribution in the tails. Such distribution is called "fat-tailed" because of large numbers of residuals far on the left and far on the right from the sample average.

Such "fat-tailed" distribution of residuals results not only in distorted values of some regression statistics, but also in the large numbers of observations in which

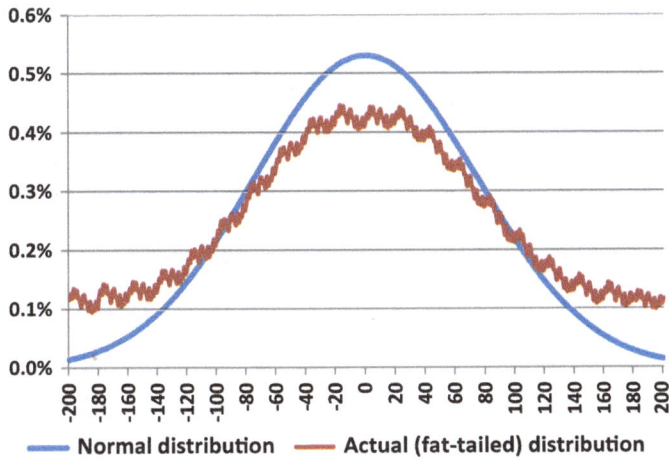

Fig. 4.2 Example of "fat-tailed" distribution of residuals. Source: Authors

the fitted values of dependent variable deviate significantly from the actual values, reducing the forecasting usefulness of the model.

4.3.4   Tests for Normality of Residuals' Distribution

Before testing the normality of the residuals, we need to compute our model residuals. Table 4.2 presents the computation.

There are many tests for checking the assumption about the normality of the distribution. However, we will present only two of them, which are easily applicable with spreadsheets

- Hellwig test for normality of distribution.
- Jarque–Bera test for normality of distribution.

4.3.5   Hellwig Test for Normality of Distribution

The Hellwig test for the normality of residuals distribution was designed by Prof. Zdzislaw Hellwig from Wroclaw University of Economics in Poland (Hellwig 1960). The test can be applied when the number of observations (residuals) is relatively small (less than 30). The procedure of this test runs as follows (Table 4.3).

To check the normality of the distribution of residuals in our model by means of the Hellwig test, the first thing is to sort all the residuals in order of their increasing values. Then we standardize the residuals on the basis of their arithmetic average and their standard deviation. The next step is to obtain the normal distribution

Table 4.2 Computation of the fitted values of the dependent variable as well as the residuals of the model

Period	COGS$_t$ (actual values of dependent variable)	NS$_t$	PRM$_t$	NPOS$_t$	$\widehat{COGS}_t$ (fitted values of dependent variable)	Residuals ($e_t = COGS_t - \widehat{COGS}_t$)
	1	2	3	4	5[a]	6 (1–5)
I q. 2008	8.700	11.200	100	2	8.878,78	−178,78
II q. 2008	10.525	13.400	105	2	10.515,20	9,80
III q. 2008	10.235	12.700	107	3	10.184,67	50,33
IV q. 2008	11.750	15.000	110	3	11.713,01	36,99
I q. 2009	11.145	14.500	109	5	11.144,25	0,75
II q. 2009	12.235	16.900	107	4	12.422,76	−187,76
III q. 2009	11.925	16.400	105	5	11.876,08	48,92
IV q. 2009	11.420	16.000	104	6	11.467,09	−47,09
I q. 2010	12.530	17.500	106	5	12.571,41	−41,41
II q. 2010	12.240	15.800	108	3	11.995,29	244,71
III q. 2010	14.155	19.000	111	2	14.129,52	25,48
IV q. 2010	15.165	20.300	113	3	14.914,28	250,72
I q. 2011	14.275	18.500	115	2	14.178,38	96,62
II q. 2011	14.575	19.600	115	1	14.895,80	−320,80
III q. 2011	15.765	21.900	113	3	15.806,51	−41,51
IV q. 2011	15.050	21.000	110	1	15.266,89	−216,89
I q. 2012	14.655	20.000	111	0	14.895,18	−240,18
II q. 2012	14.330	19.000	110	0	14.255,62	74,38
III q. 2012	14.800	20.000	108	0	14.649,42	150,58
IV q. 2012	14.295	19.000	107	0	14.009,86	285,14

Source: Authors

[a]Computed according to the obtained model, $COGS_t = -5.350{,}86 + 0{,}56\ NS_t + 81{,}92\ PRM_t - 104{,}01\ NPOS_t$

Table 4.3 The procedure of Hellwig test (for the normality of residuals' distribution)

Step 1	Formulate the hypotheses: $H_0 : F(e_t) \cong F(N)$ (residuals are normally distributed) $H_1 : F(e_t) \ncong F(N)$ (residuals are not normally distributed) where: e_t—vector of the regression residuals, $F(e_t)$—actual (empirical) distribution of residuals, $F(N)$—theoretical normal distribution of residuals.
Step 2	Sort all the residuals in order of their increasing values.
Step 3	Standardize the residuals: $e'_i = \frac{e_i - \bar{e}}{S}$ where S denotes the standard deviation of residuals: $S = \sqrt{\dfrac{1}{n} \sum_{i=1}^{n} (e_i - \bar{e})^2}$, e_i—i-th residual, $\bar{e}$—arithmetic mean of all the residuals.
Step 4	Derive the normal distribution frequency for each standardized residual, from the table of normal distribution or from Excel.
Step 5	Create n "cells," by dividing the range [0, 1] into n sub-ranges (cells) of equal length, when n denotes the number of residuals in the model.
Step 6	Compute the number of standardized residuals falling into the individual cells and then count the number of empty cells (i.e., the cells into which any standardized residual have fallen), denoted as k_e.
Step 7	Derive the critical Hellwig-test values of k_1 and k_2, for n residuals (observations), and for the assumed level of significance α.
Step 8	Compare k_e with k_1 and k_2. If $k_1 \leq k_e \leq k_2$, then H_0 is not rejected, which means that the residuals' distribution is close to normal. If $k_e < k_1$ or $k_e > k_2$, then H_0 is rejected and that the regression' residuals are not normally distributed.

Source: Authors

frequency from the table of normal distribution (or from Excel). We will show the results of these three steps in the following table (Table 4.4).

The examples of the computations of the standardized residuals are as follows:

- For the first sorted residual (equaling -320.88):

$$e'_{1s} = \frac{e_1 - \bar{e}}{S} = \frac{-320.88 - 0}{166,88} \approx -1,923$$

- For the last sorted residual (equaling 285.14):

Table 4.4 The results of the residuals sorting, standardization, and derivation of the normal distribution frequencies

Residuals $(e_t = COGS_t - \widehat{COGS_t})$	Sorted residuals e_{ts}	Standardized residuals e'_{ts}	Normal distribution frequency for standardized residuals $F(e_{ts})^a$
−178,78	−320,80	−1,922	0,027
9,80	−240,18	−1,439	0,075
50,33	−216,89	−1,300	0,097
36,99	−187,76	−1,125	0,130
0,75	−178,78	−1,071	0,142
−187,76	−47,09	−0,282	0,389
48,92	−41,51	−0,249	0,402
−47,09	−41,41	−0,248	0,402
−41,41	0,75	0,005	0,502
244,71	9,80	0,059	0,523
25,48	25,48	0,153	0,561
250,72	36,99	0,222	0,588
96,62	48,92	0,293	0,615
−320,80	50,33	0,302	0,619
−41,51	74,38	0,446	0,672
−216,89	96,62	0,579	0,719
−240,18	150,58	0,902	0,817
74,38	244,71	1,466	0,929
150,58	250,72	1,502	0,934
285,14	285,14	1,709	0,956
Arithmetic mean ($\bar{e}$)	0	0	–
Std. deviation (S)	166,88	1	–

Source: Authors

[a]The values derived from the table for normal distribution frequency or from the spreadsheet

$$e'_{1s} = \frac{e_1 - \bar{e}}{S} = \frac{285.14 - 0}{166,88} \approx 1,709$$

In the next step, we create $n = 20$ cells, by dividing the range [0, 1] into n equal sub-ranges. For $n = 20$, it gives the individual cell's width to be $0,050 = 1/20$. Then, on the basis of the normal distribution frequencies appointed for the standardized residuals, we compute the number of residuals falling into each individual cell. Finally, we compute the value of k_e, which is the number of empty cells. All these three steps are presented in Table 4.5.

As the above data show, the number of empty cells $k_e = 7$. Then we obtain the critical Hellwig test values of k_1 and k_2, for $n = 20$ observations and for the assumed level of statistical significance α (we assume $\propto = 0,05$). In the appropriate table (found in the Appendix A4), we can find that $k_1 = 4$ and $k_2 = 9$.

Table 4.5 The computation of the number of empty cells for Hellwig test

Cell	Number of residuals falling into the cell[a]
<0; 0,05)	1
<0,05; 0,10)	2
<0,10; 0,15)	2
<0,15; 0,20)	0
<0,20; 0,25)	0
<0,25; 0,30)	0
<0,30; 0,35)	0
<0,35; 0,40)	1
<0,40; 0,45)	2
<0,45; 0,50)	0
<0,50; 0,55)	2
<0,55; 0,60)	2
<0,60; 0,65)	2
<0,65; 0,70)	1
<0,70; 0,75)	1
<0,75; 0,80)	0
<0,80; 0,85)	1
<0,85; 0,90)	0
<0,90; 0,95)	2
<0,95; 1,00>	1
The number of empty cells (k_e)	7

Source: Authors
[a]On the basis of the data from the last column of Table 4.4

Since $k_1 \leq k_e \leq k_2$, H_0 is not rejected, which means that no statistical evidence was found against the normality of the residuals.

4.3.6   Jarque–Bera Test for Normality of Distribution

The Jarque–Bera test (*JB* test) constitutes an alternative tool for testing the normality of residuals distribution (Jarque and Bera 1987). However, in contrast to the Hellwig test, which was recommended for small samples (fewer than 30 observations, as a rule of thumb), the *JB* test should be applied only for relatively large samples. Both tests are, therefore, complementary, not substitutes for each other.

Jarque–Bera test is based on comparing the empirical asymmetry and kurtosis measures computed for the tested sample of residuals with the analogical measures under the normal distribution. The procedure of the test runs as follows (Table 4.6).

Even though the Jarque–Bera test is applicable in case of the large samples, for illustrative purposes we will apply this test to our model, in which the number of observations and residuals is only 20.

Table 4.6 The procedure of Jarque–Bera test (for the normality of residuals distribution)

Step 1	Formulate the hypotheses: $H_0 : F(e_t) \cong F(N)$ (residuals are normally distributed) $H_1 : F(e_t) \ncong F(N)$ (residuals are not distributed) where: e_t—vector of the regression' residuals, $F(e_t)$—actual (empirical) distribution of residuals, $F(N)$—theoretical (normal) distribution of residuals.
Step 2	For the series of tested residuals compute the following: $B_1 = A^2$ where: $A = \frac{M_3}{S^3}, \quad M_3 = \frac{1}{n} \sum_{i=1}^{n} e_i^3, \quad S = \sqrt{\frac{1}{n} \sum_{i=1}^{n} e_i^2},$ n—the number of tested residuals.
Step 3	For the same series of tested residuals compute the following: $B_2 = \frac{M_4}{S^4},$ where: $M_4 = \frac{1}{n} \sum_{i=1}^{n} e_i^4.$
Step 4	For the calculated values of B_1 and B_2 compute the empirical value of Jarque–Bera statistic (denoted as JB_e), as follows: $JB_e = n \left(\frac{1}{6} B_1 + \frac{1}{24} (B_2 - 3)^2 \right)$
Step 5	Derive the critical value of JB statistic (denoted as JB^*), from Chi-squared distribution table, for 2 degrees of freedom and for the assumed level of statistical significance α (usually assumed at 5%).
Step 6	Compare JB_e with JB^*. If $JB_e \leq JB^*$, then H_0 is not rejected, which means that the residuals' distribution is close to normal. If $JB_e \geq JB^*$, then H_0 is rejected (and H_1 is true), which means that the regression' residuals are not normally distributed.

Source: Authors

The first thing to do is to compute the values of B_1 and B_2. However, before we do this, we need to calculate the values of M_3, M_4, S, and A. We show the results of all these computations in the following table (Table 4.7).

The following computations are as follows. For $n = 20$:

$$S = \sqrt{\frac{1}{n} \sum_{i=1}^{n} e_i^2} = \sqrt{\frac{529.144, 90}{20}} \approx 162, 66,$$

$$M_3 = \frac{1}{n} \sum_{i=1}^{n} e_i^3 = \frac{-11.014.148, 52}{20} \approx -550.707, 43,$$

Table 4.7 The computations for Jarque–Bera test of normality of residuals' distribution

Residuals ($e_t = COGS_t - \widehat{COGS_t}$)	e_t^2	e_t^3	e_t^4
−178,78	31.962,27	−5.714.211,72	1.021.586.402,80
9,80	96,12	942,32	9.238,37
50,33	2.532,61	127.454,03	6.414.134,96
36,99	1.367,97	50.595,96	1.871.347,37
0,75	0,57	0,43	0,32
−187,76	35.253,34	−6.619.123,23	1.242.798.218,57
48,92	2.392,86	117.050,86	5.725.755,81
−47,09	2.217,71	−104.437,45	4.918.224,23
−41,41	1.714,81	−71.010,49	2.940.559,56
244,71	59.884,85	14.654.648,35	3.586.194.738,70
25,48	649,42	16.549,82	421.752,61
250,72	62.860,90	15.760.531,33	3.951.492.294,09
96,62	9.335,83	902.047,03	87.157.662,56
−320,80	102.910,94	−33.013.556,90	10.590.661.582,67
−41,51	1.722,80	−71.507,71	2.968.045,05
−216,89	47.042,50	−10.203.181,00	2.212.996.808,71
−240,18	57.688,60	−13.855.906,54	3.327.974.017,01
74,38	5.532,13	411.470,87	30.604.511,80
150,58	22.673,61	3.414.137,87	514.092.663,64
285,14	81.305,06	23.183.357,65	6.610.512.235,78
Σ	529.144,90	−11.014.148,52	33.201.340.194,61

Source: Authors

$$M_4 = \frac{1}{n} \sum_{i=1}^{n} e_i^4 = \frac{33.201.340.194,61}{20} \approx 1.660.067.009,73,$$

$$A = \frac{M_3}{S^3} = \frac{-550.707,43}{(162,66)^3} \approx -0,1280,$$

$$B_1 = A^2 = (-0,1280)^2 \approx 0,0164,$$

$$B_2 = \frac{M_4}{S^4} = \frac{1.660.067.009,73}{(162,66)^4} \approx 2,3716.$$

Now that we have computed the values of B_1 and B_2, we can easily obtain the value of empirical Jarque–Bera statistic (JB_e), as follows:

$$JB_e = n\left(\frac{1}{6}B_1 + \frac{1}{24}(B_2 - 3)^2\right) = 20\left(\frac{1}{6}0,0164 + \frac{1}{24}(2,3716 - 3)^2\right) = 0,3837.$$

The next step is to derive the critical value of JB statistic (JB^*), from Chi-squared distribution table, for 2 degrees of freedom and for the assumed level of statistical significance α (which we will assume at 5%).

$$JB^* = 5,99.$$

Given that in our model $JB_e \leq JB^*$, H_0 is not rejected, which means that the residuals' distribution can be assumed to be normal.

4.4   Testing the Autocorrelation of Regression Residuals

4.4.1   Nature and Relevance of Autocorrelation

In a correctly estimated model, the residuals should be chaotic and not related to each other. Therefore, the presence of significant autocorrelation of residuals signals that something is wrong with the estimated model. Autocorrelation is observed more frequently in time-series models (Cochrane and Orcutt 1949).

The presence of autocorrelation brings the following problems:

- The overstatement or understatement of standard errors computed for the regression coefficients,
- The tendency of the model to systematically understate or overstate the fitted values and forecasts of the dependent variable.

Among the most common causes of autocorrelation are:

- Incorrect functional form of the estimated regression (e.g., linear instead of nonlinear).
- Existence of significant missing explanatory variables with strong relationships with the dependent variable which were not included in the model.
- Structural changes occurring within the estimated relationships.

4.4.2   Illustration of Incorrect Functional Form as the Cause of Autocorrelation

Suppose that we are tracking the relationship between a dependent variable Y and an explanatory variable X. The actual observations as well as the linear regression of Y against X are shown in the following chart (Fig. 4.3).

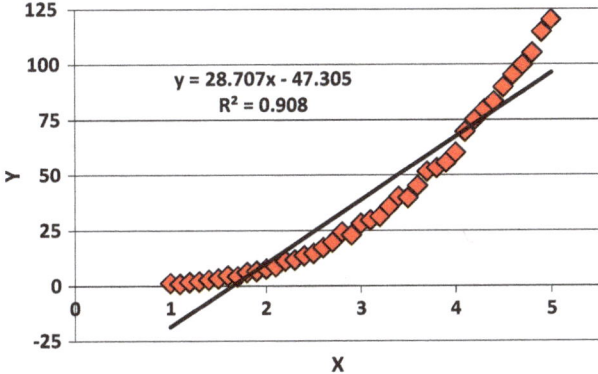

Fig. 4.3 The actual nonlinear relationship between Y and X and linear regression of Y against X.
Source: Authors

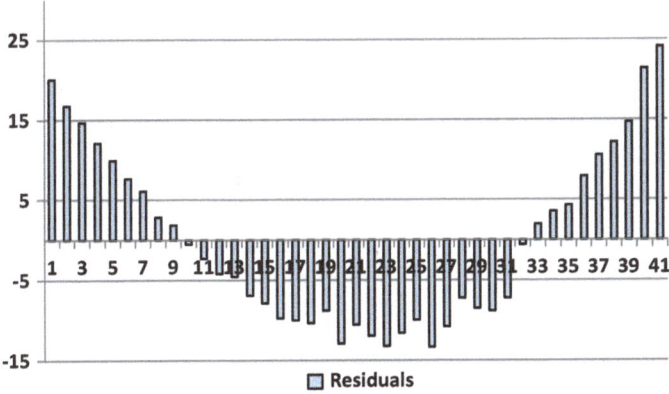

Fig. 4.4 The residuals from the linear regression of Y against X. Source: Authors

The above figure shows the incorrectly assumed linear form of the regression of Y against X. The following figure presents the residuals of the relationship between Y and X from the linear regression.

The pattern of residuals presented in Fig. 4.4 clearly cannot be described as chaotic or noisy, showing that the residuals from individual observations are related to each other. This is confirmed by the following scatterplot, showing the relationship between current residuals and lagged (by one observation) residuals. This is the first-order autocorrelation of residuals (Fig. 4.5).

The chart shows that the residuals are highly correlated with the residuals lagged by one observation. The correlation coefficient is very large, 0,97, which confirms the strong first-order autocorrelation. In this case, the reason for such strong autocorrelation is the incorrect functional form of the investigated relationship.

Fig. 4.5 The first-order
autocorrelation of residuals
from the linear regression of
Y against X. Source:
Authors

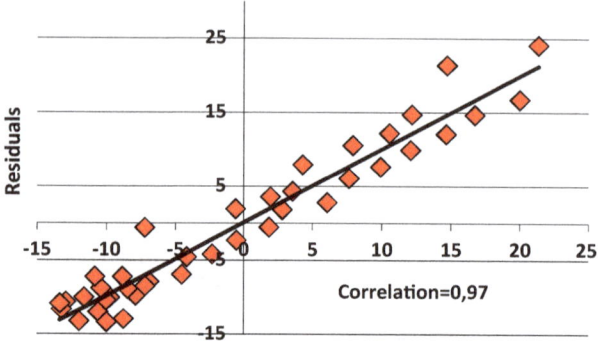

Fig. 4.6 The actual
(nonlinear) relationship
between Y and X and power
regression of Y against X.
Source: Authors

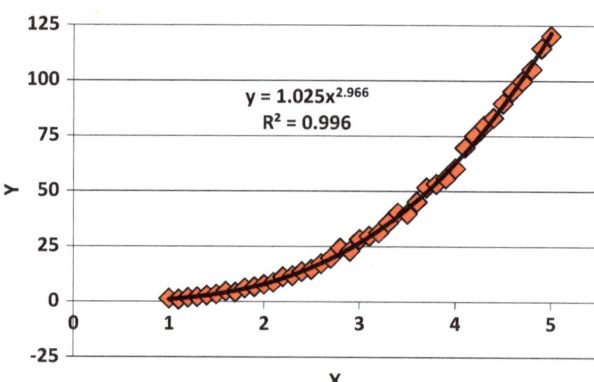

The following figure shows the change of the functional form from linear regression
to power regression (Fig. 4.6).

As can be seen in the above figure, the power form of the regression of Y against
X results in much better fit to the actual data. This model better captures the actual
relationship (it has a larger value of R-squared than in the case of linear regression),
but also results in much more chaotic residuals, shown in the following figure
(Fig. 4.7).

The much weaker first-order autocorrelation of residuals, than in case of linear
regression, is confirmed by the following scatterplot (Fig. 4.8).

As we can see, the residuals are now insignificantly correlated with the residuals
lagged by one observation. Now the correlation coefficient is low (0,18) and
statistically insignificant (at 5% significance level), which confirms the lack of
first-order autocorrelation.

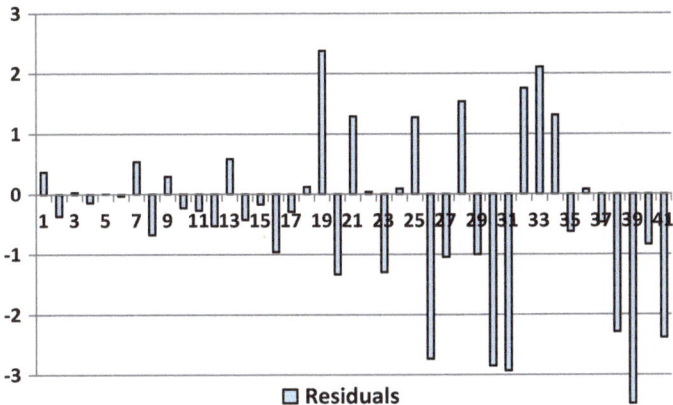

Fig. 4.7 The residuals from the power regression of Y against X. Source: Authors

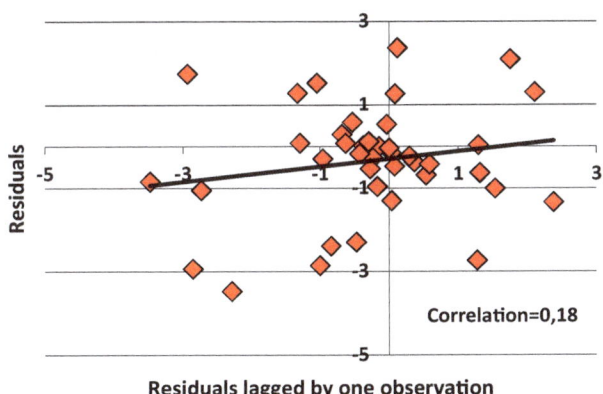

Fig. 4.8 The first-order autocorrelation of residuals from the power regression of Y against X. Source: Authors

4.4.3   Illustration of Missing Explanatory Variable as the Cause of Autocorrelation

Suppose that we are building the regression explaining Y as a function of some explanatory variables. We hypothesize that $X1$ is the only driver of changes of Y, while in reality Y is driven by both $X1$ and $X2$. If we forget about including both $X1$ and $X2$ in the set of explanatory variables, then we can obtain the regression with

Table 4.8 Actual data of Y, $X1$, and $X2$ and residuals from the regression of Y against $X1$ only

Y	X1	X2	Fitted Y	Residuals e_t	First lag e_{t-1}
55	1	49	42,7	12,3	–
49	2	50	43,7	5,3	12,3
47	3	42	44,6	2,4	5,3
42	4	35	45,6	−3,6	2,4
41	5	33	46,5	−5,5	−3,6
45	6	37	47,5	−2,5	−5,5
50	7	48	48,4	1,6	−2,5
55	8	47	49,4	5,6	1,6
51	9	44	50,4	0,6	5,6
43	10	35	51,3	−8,3	0,6
44	11	34	52,3	−8,3	−8,3
48	12	35	53,2	−5,2	−8,3
53	13	41	54,2	−1,2	−5,2
55	14	45	55,1	−0,1	−1,2
58	15	44	56,1	1,9	−0,1
53	16	38	57,0	−4,0	1,9
53	17	30	58,0	−5,0	−4,0
55	18	36	59,0	−4,0	−5,0
60	19	42	59,9	0,1	−4,0
67	20	53	60,9	6,1	0,1
72	21	49	61,8	10,2	6,1
68	22	44	62,8	5,2	10,2
65	23	38	63,7	1,3	5,2
61	24	35	64,7	−3,7	1,3
63	25	40	65,7	−2,7	−3,7
70	26	46	66,6	3,4	−2,7
75	27	48	67,6	7,4	3,4
70	28	41	68,5	1,5	7,4
63	29	34	69,5	−6,5	1,5
62	30	32	70,4	−8,4	−6,5
70	31	37	71,4	−1,4	−8,4
76	32	43	72,3	3,7	−1,4
77	33	46	73,3	3,7	3,7
72	34	44	74,3	−2,3	3,7
72	35	37	75,2	−3,2	−2,3
67	36	33	76,2	−9,2	−3,2
70	37	33	77,1	−7,1	−9,2
80	38	39	78,1	1,9	−7,1
87	39	44	79,0	8,0	1,9
90	40	54	80,0	10,0	8,0

Source: Authors

significantly autocorrelated residuals. Table 4.8 presents the actual data as well as the residuals from the linear regression of Y against $X1$ only. Table 4.9 presents the

Table 4.9 Actual data of Y, $X1$, and $X2$ and residuals from the regression of Y against both $X1$ and $X2$

Y	$X1$	$X2$	Fitted Y	Residuals e_t	First lag e_{t-1}
55	1	49	48,9	6,1	–
49	2	50	50,6	−1,6	6,1
47	3	42	45,2	1,8	−1,6
42	4	35	40,6	1,4	1,8
41	5	33	39,9	1,1	1,4
45	6	37	44,1	0,9	1,1
50	7	48	53,9	−3,9	0,9
55	8	47	54,1	0,9	−3,9
51	9	44	52,6	−1,6	0,9
43	10	35	46,4	−3,4	−1,6
44	11	34	46,6	−2,6	−3,4
48	12	35	48,4	−0,4	−2,6
53	13	41	54,1	−1,1	−0,4
55	14	45	58,3	−3,3	−1,1
58	15	44	58,5	−0,5	−3,3
53	16	38	54,7	−1,7	−0,5
53	17	30	49,2	3,8	−1,7
55	18	36	55,0	0,0	3,8
60	19	42	60,8	−0,8	0,0
67	20	53	70,6	−3,6	−0,8
72	21	49	68,3	3,7	−3,6
68	22	44	65,3	2,7	3,7
65	23	38	61,5	3,5	2,7
61	24	35	60,1	0,9	3,5
63	25	40	65,0	−2,0	0,9
70	26	46	70,8	−0,8	−2,0
75	27	48	73,4	1,6	−0,8
70	28	41	68,8	1,2	1,6
63	29	34	64,1	−1,1	1,2
62	30	32	63,5	−1,5	−1,1
70	31	37	68,5	1,5	−1,5
76	32	43	74,3	1,7	1,5
77	33	46	77,6	−0,6	1,7
72	34	44	77,0	−5,0	−0,6
72	35	37	72,4	−0,4	−5,0
67	36	33	70,1	−3,1	−0,4
70	37	33	71,1	−1,1	−3,1
80	38	39	76,9	3,1	−1,1
87	39	44	81,9	5,1	3,1
90	40	54	90,9	−0,9	5,1

Source: Authors

same actual data as well as the residuals from the linear regression of Y against both $X1$ and $X2$.

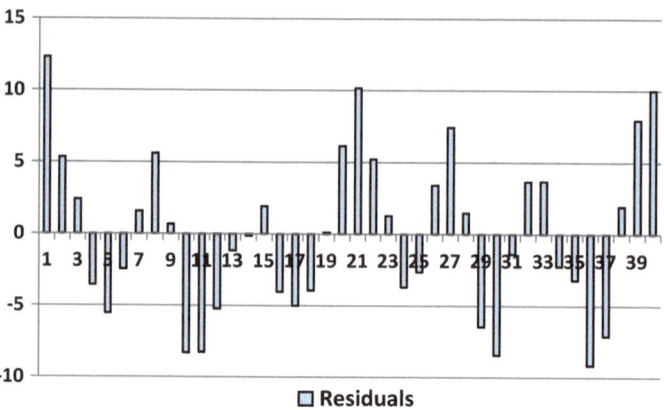

Fig. 4.9 The residuals from the regression of Y against X1 only. Source: Authors

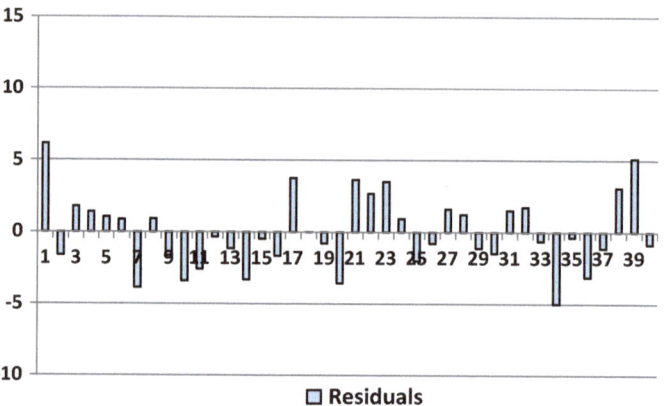

Fig. 4.10 The residuals from the regression of Y against both X1 and X2. Source: Authors

Figures 4.9 and 4.10 present the residuals from both regressions. Figures 4.11 and 4.12 present the first-order autocorrelation of the residuals obtained from these regressions. As can be inferred from the comparison of Fig. 4.9 with Fig. 4.10, and particularly from comparing Fig. 4.11 with Fig. 4.12, residuals from the regression of Y against both X1 and X2 are more chaotic and less autocorrelated than the residuals from regressing Y against only X1.

The coefficient of first-order autocorrelation is quite large (0,60) and statistically significant at the 5% significance level when Y is regressed against X1 only, while coefficient of first-order autocorrelation is small (0,13) and statistically insignificant when Y is regressed against both X1 and X2. This confirms that missing X2 in regression of Y results in significant autocorrelation of residuals.

Fig. 4.11 The first-order autocorrelation of residuals from the regression of Y against $X1$ only. Source: Authors

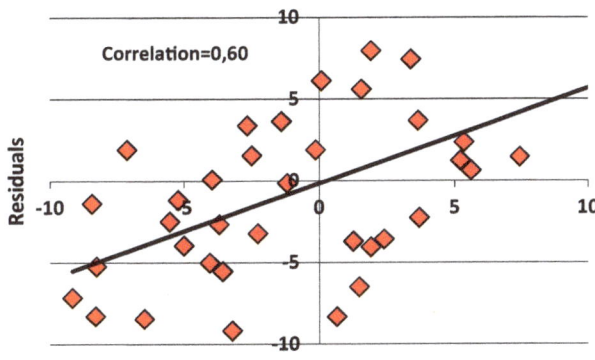

Fig. 4.12 The first-order autocorrelation of residuals from the regression of Y against both $X1$ and $X2$. *Source: Authors*

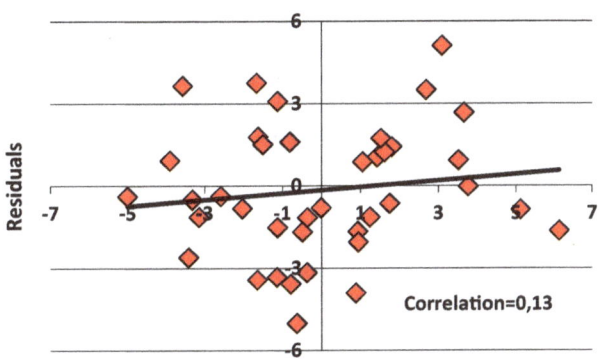

4.4.4 Illustration of Distorting Impact of Autocorrelation on t-Statistics

We will illustrate the distorting effect of autocorrelation on the basis of the same data which were used in the above regression of Y against $X1$ and $X2$. Table 4.10 presents the basic parameters of the regression of Y against $X1$ only, and Table 4.11 presents the analogous parameters of regressing Y against both $X1$ and $X2$.

As can be seen, the autocorrelation present in the reduced regression (i.e., Y against $X1$ only) caused the slope coefficient to have an understated t-Statistic of significance (only 12,347). In the full regression of Y against both $X1$ and $X2$ not only the regression residuals have much lower absolute values, on average, but also the t-Statistic of significance of $X1$ is more than twice larger than in the reduced model (i.e., 27,190 vs. 12,347).

The above example corroborates that it is very important to control for the autocorrelation of residuals when building and interpreting econometric models. In the case of autocorrelation, some explanatory variables may seem statistically

Table 4.10 Results of regressing Y against $X1$ only

Variable	Coefficient	t-Statistic
Intercept	41,75	22,911
$X1$	0,96	12,347
R-squared: 0,800		

Source: Authors

Table 4.11 Results of regressing Y against both $X1$ and $X2$

Variable	Coefficient	t-Statistic
Intercept	8,66	2,972
$X1$	0,97	27,190
$X2$	0,80	11,867
R-squared: 0,958		

Source: Authors

insignificant, while actually they may have significant relationships with the dependent variable. Note also that the autocorrelation of residuals in the reduced regression only brought about the understatement of the t-Statistic for the slope parameter. The slope coefficient itself was not significantly distorted (coefficient at $X1$ equals 0,96 in the reduced regression as compared to 0,97 in the full regression—the difference of only about 1%).

4.4.5 F-test (Fisher–Snedecor Test) for the Autocorrelation of Residuals

If regression residuals are autocorrelated, then there exists a linear relationship between current (unlagged) residuals and lagged residuals. Therefore, we can apply the simple test for autocorrelation, based on multiple regression, in which our residuals constitute the dependent variable and lagged residuals (up to the lag-order preset, usually quite arbitrarily, by the researcher) constitute the set of explanatory variables (Charemza and Deadman 1992).

We know from the F-test for the statistical significance of the whole set of explanatory variables that if at least one explanatory variable is statistically significant, then the F-test points to the statistical significance of the whole regression. So, if the F-test for such regression of residuals against lagged residuals indicates that this whole regression is statistically significant, then we receive the information that some autocorrelation of residuals exists. The procedure of the F-test for autocorrelation runs as follows (Table 4.12).

In checking the presence of residuals autocorrelation by means of the F-test, the first thing to do is to prepare the data by lagging the residuals. The subjective and arbitrary issue here is to specify the maximum considered order of autocorrelation (k). In terms of data with frequency higher than annual, as a rule of thumb the researchers often set the maximum lag as some (arbitrary) multiple of order of

Table 4.12 Procedure for the F-test for the autocorrelation of residuals

Step 1	Formulate the hypotheses:
	$H_0 : r_{e_t,e_{t-1}} = r_{e_t,e_{t-2}} = \ldots = r_{e_t,e_{t-k}} = 0$ (residuals are not autocorrelated)
	$H_1 : r_{e_t,e_{t-1}} \neq 0 \vee r_{e_t,e_{t-2}} \neq 0 \vee \ldots \vee r_{e_t,e_{t-k}} \neq 0$ (residuals are autocorrelated)
	where:
	e_t—vector of the regression' residuals,
	$r_{e_t,e_{t-i}}$—autocorrelation between unlagged residuals and residuals lagged by i-th order,
	k—maximum assumed lag of residuals (maximum assumed order of autocorrelation).
Step 2	Regress the residuals against the lagged residuals, i.e., estimate the regression of the following form:
	$e_t = \alpha_0 + \alpha_1 e_{t-1} + \alpha_2 e_{t-2} + \ldots + \alpha_k e_{t-k}$
Step 3	For the above regression of residuals against lagged residuals, compute the empirical statistic F according to the following formula:
	$F = \frac{R^2}{1-R^2} \frac{n-(k+1)}{k}$,
	where:
	R^2—coefficient of determination (R-squared) of the regression of residuals against lagged residuals,
	n—number of observations on which the regression of residuals was estimated,
	k—number of explanatory variables in the regression.
Step 4	Obtain the critical (theoretical) value F^* from the Fisher–Snedecor table, for $m_1 = k$, $m_2 = n - (k+1)$ and for the assumed significance level α.
Step 5	Compare the value of the empirical F with the critical F^*.
	If $F \leq F^*$, then H_0 is not rejected, which means that the data do not support a conclusion of the existence of autocorrelation of up to k-th order.
	If $F > F^*$, then H_0 is rejected, which means that the data evidence some autocorrelation of the regression' residuals.

Source: Authors

seasonality. This means that in terms of monthly data usually lags of up to 12 or 24 observations are considered and in terms of quarterly data usually lags of up to 4 or 8 or 12 observations are considered.

In terms of annual-frequency data, usually the researcher's experience and knowledge as to the time structure of the processes modeled is being applied. Here, we have quarterly data and we will consider the autocorrelation of up to fourth quarter only, because we are restrained by our relatively small sample of 20 residuals, which for F-test is reduced even more by the lags. The data prepared for the regression of residuals against lagged residuals are shown below (Table 4.13).

We need to regress the residuals against the lagged residuals. This regression has the following form:

$$e_t = \alpha_0 + \alpha_1 e_{t-1} + \alpha_2 e_{t-2} + \alpha_3 e_{t-3} + \alpha_4 e_{t-4}.$$

The results of this regression are presented in Table 4.14.

Although our model, which is now tested for autocorrelation, was estimated on the basis of 20 observations, in testing the autocorrelation of up to fourth order we have only 16 observations, because we lost 4 observations by lagging residuals.

Table 4.13 The residuals and lagged residuals up to the fourth order from our model

Residuals e_t	First lag e_{t-1}	Second lag e_{t-2}	Third lag e_{t-3}	Fourth lag e_{t-4}
−178,78	−	−	−	−
9,80	−178,78	−	−	−
50,33	9,80	−178,78	−	−
36,99	50,33	9,80	−178,78	−
0,75	36,99	50,33	9,80	−178,78
−187,76	0,75	36,99	50,33	9,80
48,92	−187,76	0,75	36,99	50,33
−47,09	48,92	−187,76	0,75	36,99
−41,41	−47,09	48,92	−187,76	0,75
244,71	−41,41	−47,09	48,92	−187,76
25,48	244,71	−41,41	−47,09	48,92
250,72	25,48	244,71	−41,41	−47,09
96,62	250,72	25,48	244,71	−41,41
−320,80	96,62	250,72	25,48	244,71
−41,51	−320,80	96,62	250,72	25,48
−216,89	−41,51	−320,80	96,62	250,72
−240,18	−216,89	−41,51	−320,80	96,62
74,38	−240,18	−216,89	−41,51	−320,80
150,58	74,38	−240,18	−216,89	−41,51
285,14	150,58	74,38	−240,18	−216,89

Source: Authors

This means that for the F-test for autocorrelation $n = 16$, there are four explanatory variables (which means that $k = 4$) and the coefficient of determination $R^2 = 0,608$. Putting these numbers into the formula for the empirical statistic F produces the following results:

$$F = \frac{R^2}{1 - R^2} \frac{n - (k + 1)}{k} = \frac{0,608}{1 - 0,608} \frac{16 - (4 + 1)}{4} = 4,26.$$

Now we need to obtain the critical value of F^*. For $n = 16$ and $k = 4$ we obtain:

$$m_1 = k = 4,$$

$$m_2 = n - (k + 1) = 16 - (4 + 1) = 11.$$

For the level of statistical significance set at $\alpha = 0,05$, and computed values of m_1 and m_2, we can obtain the critical value of F^* from the Fisher–Snedecor table, which equals 3,36.

Finally, we compare the value of the empirical statistic F with its critical value F^*. In our case $F > F^*$, which means that H_0 is rejected, and we conclude that there is some autocorrelation of the residuals in our model.

Table 4.14 Results of regressing residuals against lagged (up to fourth order) residuals

Variable	Coefficient	t-Statistic
Intercept	−5,83	−0,173
First lag e_{t-1}	0,27	1,282
Second lag e_{t-2}	0,04	0,197
Third lag e_{t-3}	0,02	0,107
Fourth lag e_{t-4}	−0,88	−3,936
R-squared: 0,608		

Source: Authors

Looking at t-Statistics of individual slope coefficients in our regression of residuals against lagged residuals, we can suppose that the autocorrelation is negative and is of fourth order, meaning that the residuals are negatively correlated with the residuals lagged by fourth quarters. This is suggested by relatively high absolute value of t-Statistic computed for the fourth lag e_{t-4}.

4.4.6   Box–Pearce Test for the Autocorrelation of Residuals

If regression residuals are autocorrelated, then the current unlagged residuals and lagged residuals are correlated. Therefore, we can apply the simple test for auto-correlation, based on correlation coefficients between residuals and lagged residuals. For example, if we consider autocorrelation of up to fourth order, then we can compute four correlation coefficients, between unlagged residuals and each of the individual four lags. If at least one of these correlation coefficients turns out to be statistically significant, then we receive the information that some autocorrelation of residuals exists. One of the tests for autocorrelation, based on coefficients of correlation between residuals, is Box–Pearce test which runs as follows (Box and Pearce 1970) (Table 4.15).

Let's check the presence of autocorrelation of residuals in our model by means of the Box–Pearce test. Similarly as in F-test, the first thing to do is to prepare the data by lagging the residuals. Again, we need to subjectively specify the maximum considered order of autocorrelation (k). We have quarterly data, so we will consider the autocorrelation of up to fourth quarter. The data prepared for computing the correlations between residuals and lagged residuals as well as the obtained correlations are shown in Table 4.16.

Although our model (which is now tested for autocorrelation) was estimated on the basis of 20 observations, in testing the autocorrelation of up to fourth order we again have only 16 observations, because we lost 4 observations by lagging residuals. This means that for the Box–Pearce test for autocorrelation, $n = 16$. Substituting the appropriate numbers into the formula for the empirical statistic Q (k) produces the following results:

Table 4.15 The procedure of Box–Pearce test for the autocorrelation of residuals

Step 1	Formulate the hypotheses: $H_0 : r_{e_t,e_{t-1}} = r_{e_t,e_{t-2}} = \ldots = r_{e_t,e_{t-k}} = 0$ (residuals are not autocorrelated) $H_1 : r_{e_t,e_{t-1}} \neq 0 \vee r_{e_t,e_{t-2}} \neq 0 \vee \ldots \vee r_{e_t,e_{t-k}} \neq 0$ (residuals are autocorrelated) where: e_t—vector of the regression' residuals, $r_{e_t,e_{t-i}}$—autocorrelation between unlagged residuals and residuals lagged by i-th order, k—maximum assumed lag of residuals (maximum assumed order of autocorrelation).
Step 2	Compute k coefficients of correlation between unlagged residuals and lagged residuals, i.e., compute the following coefficients: $r_{e_t,e_{t-1}}$—coefficient of correlation between unlagged residuals and residuals lagged by first order (first-order autocorrelation), $r_{e_t,e_{t-2}}$—coefficient of correlation between unlagged residuals and residuals lagged by second order (second-order autocorrelation), $\ldots\ldots\ldots\ldots$, $r_{e_t,e_{t-k}}$- coefficient of correlation between unlagged residuals and residuals lagged by k-th order (k-order autocorrelation).
Step 3	On the basis of k coefficients of correlation, compute the empirical statistic $Q(k)$ according to the following formula: $$Q(k) = n \sum_{i=1}^{k} r^2_{e_t,e_{t-i}},$$ where: n—number of observations on which coefficients of correlation were computed.
Step 4	Obtain the critical (theoretical) value $Q(k)^*$ from the Chi-squared distribution table, for k degrees of freedom and for assumed significance level α (usually at 5%).
Step 5	Compare empirical $Q(k)$ with critical $Q(k)^*$. If $Q(k) \leq Q(k)^*$, then H_0 is not rejected, which means that the regression residuals are not autocorrelated up to k-th order. If $Q(k) > Q(k)^*$, then H_0 is rejected (and H_1 is true), which means that there is some autocorrelation of the regression residuals.

Source: Authors

$$Q(k) = n \sum_{i=1}^{4} r^2_{e_t,e_{t-i}} = 16\left((0,217)^2 + (0,038)^2 + (-0,094)^2 + (-0,738)^2\right)$$

$$= 9,62$$

Now we obtain the critical value of $Q(k)^*$. For the level of statistical significance set at $\alpha = 0,05$ and for $k=4$, we obtain the critical value of $Q(k)^*$ from the Chi-squared table or from Excel, which equals 9,49.

Finally, we compare the value of empirical statistic $Q(k)$ with its critical value $Q(k)^*$. In our case $Q(k) > Q(k)^*$, which means that H_0 is rejected, and we conclude that there is some autocorrelation of the residuals in our model.

Looking at individual coefficients of correlation between residuals and lagged residuals, we can conclude that the autocorrelation is negative and is of fourth order, meaning that the residuals are negatively correlated with the residuals lagged

Table 4.16 The residuals from our model and coefficients of correlation between unlagged residuals and residuals lagged up to the fourth order

Residuals e_t	First lag e_{t-1}	Second lag e_{t-2}	Third lag e_{t-3}	Fourth lag e_{t-4}
−178,78	−	−	−	−
9,80	−178,78	−	−	−
50,33	9,80	−178,78	−	−
36,99	50,33	9,80	−178,78	−
0,75	36,99	50,33	9,80	−178,78
−187,76	0,75	36,99	50,33	9,80
48,92	−187,76	0,75	36,99	50,33
−47,09	48,92	−187,76	0,75	36,99
−41,41	−47,09	48,92	−187,76	0,75
244,71	−41,41	−47,09	48,92	−187,76
25,48	244,71	−41,41	−47,09	48,92
250,72	25,48	244,71	−41,41	−47,09
96,62	250,72	25,48	244,71	−41,41
−320,80	96,62	250,72	25,48	244,71
−41,51	−320,80	96,62	250,72	25,48
−216,89	−41,51	−320,80	96,62	250,72
−240,18	−216,89	−41,51	−320,80	96,62
74,38	−240,18	−216,89	−41,51	−320,80
150,58	74,38	−240,18	−216,89	−41,51
285,14	150,58	74,38	−240,18	−216,89
Coefficients of correlation	$r_{e_t,e_{t-1}}$	$r_{e_t,e_{t-2}}$	$r_{e_t,e_{t-3}}$	$r_{e_t,e_{t-4}}$
	0,217	0,038	−0,094	−0,738

Source: Authors

by fourth quarters. This is suggested by relatively large and statistically significant absolute value of $r_{e_t,e_{t-4}}$.

4.4.7   Interpretation of Autocorrelation Tests Conducted for Our Model

Both tests (*F*-test and Box–Pearce test) pointed to the existence of some autocorrelation. Results of both tests suggest that the autocorrelation is of the fourth order, because:

- In *F*-test, the fourth lag of residuals is statistically significant (given its *t*-Statistic).
- In Box–Pearce test, the coefficient of correlation between current residuals and residuals lagged by fourth quarters is statistically significant.

The presence of this autocorrelation suggests that perhaps:

- The functional form (linear) of our model is not correct.
- The model is misspecified in that sense that some important variables are missing.
- There are some structural changes of the modeled relationship occurring in our sample.

Later on we will test the correctness of functional form by means of tests for residuals symmetry and by means of the Ramsey's *RESET* test. If all the tests for the functional form suggest that the linearity assumption is valid, then it will be necessary to consider some additional variables or some structural changes in the modeled relationships.

If for some reasons the researcher does not want to change the model, in which the residuals are autocorrelated, then the model should be reestimated with the use of some method of estimation which is more immune than ordinary least squares to the distorting effects of autocorrelation, such as weighted least squares.

4.5   Testing the Heteroscedasticity of Regression Residuals

4.5.1   Nature and Relevance of Heteroscedasticity

One of the assumptions of ordinary least squares states that the variance of residuals is constant (homoscedastic). Homoscedasticity of residuals means that residuals obtained from estimated regression have comparable variance, regardless of in which part of the sample we are.

In contrast, heteroscedasticity of residuals means that the variance of residuals is not constant within a sample. For example, in time-series models it can happen that the variance of residuals is relatively small at the beginning of the sample and it is relatively large at the end of the sample. In cross-sectional models, the variance of residuals can be positively or negatively correlated with one or several explanatory variables, meaning that, for example, the variance of residuals is relatively small in case of observations with relatively small values of some explanatory variable and it is larger for observations with larger values of explanatory variable. In any case, the presence of heteroscedasticity is unwelcome and it breaks down one of the important assumptions of ordinary least squares.

The presence of heteroscedasticity of residuals brings the following problems (White 1980):

- The overstatement or understatement of standard errors computed for the regression coefficients (entailing overstatement or understatement of t-Statistics).
- The tendency of the model to generate fitted values as well as forecasts of the dependent variable with differing scope of average errors in different points within the sample.

Among the most common causes of heteroscedasticity are:

Table 4.17 Actual data (of Y, $X1$, and $X2$) and residuals from the regression of Y against $X1$ only

Y	$X1$	$X2$	Fitted Y	Residuals e_t	Absolute values of residuals $|e_t|$
7	1	1	5,5	1,5	1,5
8	2	−1	10,4	−2,4	2,4
15	3	2	15,4	−0,4	0,4
20	4	−2	20,4	−0,4	0,4
27	5	3	25,3	1,7	1,7
27	6	−3	30,3	−3,3	3,3
40	7	4	35,2	4,8	4,8
35	8	−4	40,2	−5,2	5,2
52	9	5	45,2	6,8	6,8
42	10	−5	50,1	−8,1	8,1
62	11	6	55,1	6,9	6,9
55	12	−6	60,0	−5,0	5,0
70	13	7	65,0	5,0	5,0
65	14	−7	70,0	−5,0	5,0
83	15	8	74,9	8,1	8,1
71	16	−8	79,9	−8,9	8,9
95	17	9	84,8	10,2	10,2
80	18	−9	89,8	−9,8	9,8
107	19	10	94,8	12,2	12,2
91	20	−10	99,7	−8,7	8,7

Source: Authors

- Incorrect functional form of the estimated regression.
- Existence of significant missing explanatory variables.
- Structural changes occurring within the estimated relationships.
- Changing nature of the variables included in the model.

4.5.2   Illustration of Heteroscedasticity of Residuals

Let's suppose that we are building the regression explaining Y (as dependent variable) as a function of some explanatory variables. We hypothesize that $X1$ is the only driver of changes of Y, while in reality Y is driven by both $X1$ and $X2$. If we forget about including both $X1$ and $X2$ in the set of explanatory variables, then we can obtain the regression with significantly heteroscedastic residuals. Table 4.17 presents the actual data for Y, $X1$, and $X2$ as well as the residuals from the linear regression of Y against $X1$ only. Table 4.18 presents the same actual data as well as the residuals from the linear regression of Y against both $X1$ and $X2$.

Figures 4.13 and 4.14 present the absolute values of residuals from both regressions. As can be inferred from Fig. 4.13, the absolute values of residuals from the regression of Y against $X1$ show an evident rising trend. In contrast, the absolute

Table 4.18 Actual data (of Y, $X1$, and $X2$) and residuals from the regression of Y against both $X1$ and $X2$

Y	$X1$	$X2$	Fitted Y	Residuals e_t	Absolute values of residuals $|e_t|$
7	1	1	5,7	1,3	1,3
8	2	−1	8,7	−0,7	0,7
15	3	2	16,8	−1,8	1,8
20	4	−2	17,7	2,3	2,3
27	5	3	28,0	−1,0	1,0
27	6	−3	26,8	0,2	0,2
40	7	4	39,1	0,9	0,9
35	8	−4	35,8	−0,8	0,8
52	9	5	50,3	1,7	1,7
42	10	−5	44,9	−2,9	2,9
62	11	6	61,4	0,6	0,6
55	12	−6	53,9	1,1	1,1
70	13	7	72,5	−2,5	2,5
65	14	−7	62,9	2,1	2,1
83	15	8	83,7	−0,7	0,7
71	16	−8	72,0	−1,0	1,0
95	17	9	94,8	0,2	0,2
80	18	−9	81,0	−1,0	1,0
107	19	10	105,9	1,1	1,1
91	20	−10	90,1	0,9	0,9

Source: Authors

Fig. 4.13 The absolute values of residuals from the regression of Y against $X1$ only and their relationship with the numbers of observations. Source: Authors

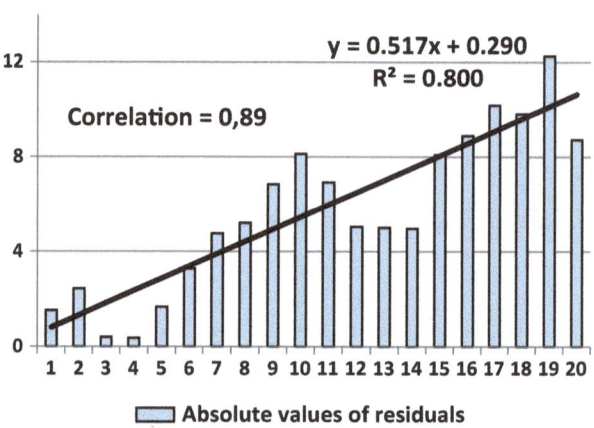

values of residuals from the regression of Y against both $X1$ and $X2$ do not show any statistically significant trends.

As we can see on both charts above, the coefficient of correlation between absolute values of residuals and numbers of observations is large (0,89) and statistically significant at the 5% significance level when Y is regressed against

Fig. 4.14 The absolute values of residuals from the regression of Y against both $X1$ and $X2$ and their relationship with the numbers of observations. Source: Authors

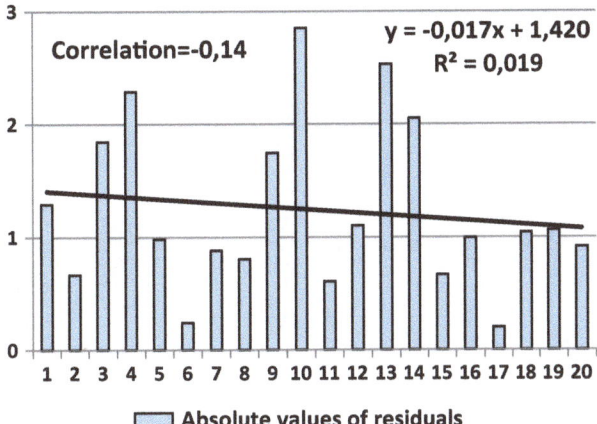

$X1$ only, while the analogous coefficient of correlation is small $(-0,14)$ and statistically insignificant when Y is regressed against both $X1$ and $X2$. This confirms that missing $X2$ in regression of Y results in significant heteroscedasticity of residuals.

It is also worth noting that the variance of $X2$ itself also shows a rising trend. If this variable with rising variability is omitted in the set of explanatory variables, while having significant relationships with the dependent variable, then the result is rising variability of regression residuals (i.e., heteroscedasticity).

4.5.3   Illustration of Distorting Impact of Heteroscedasticity on t-Statistics

We will illustrate the distorting effect of heteroscedasticity on the basis of the same data which were used in the above regression of Y against $X1$ and $X2$. Table 4.19 presents the basic parameters of the regression of Y against $X1$ only, and Table 4.20 presents the analogous parameters of regressing Y against both $X1$ and $X2$.

As can be seen, the heteroscedasticity present in the reduced regression of Y against $X1$ caused the slope coefficient to have understated t-Statistic of significance (only 18,325). In the full regression of Y against both $X1$ and $X2$, the t-Statistic of significance of $X1$ is more than four times larger than in the reduced model (i.e., 83,455 vs. 18,325).

The above example corroborates that it is very important to control for the heteroscedasticity of residuals when building and interpreting econometric models. Note also that similarly to autocorrelation, the heteroscedasticity of residuals in the reduced regression brought about only the understatement of the t-Statistic for the slope parameter. The slope coefficient itself was not significantly distorted. The coefficient at

Table 4.19 Results of
regressing Y against $X1$ only

Variable	Coefficient	t-Statistic
Intercept	0,53	0,162
$X1$	4,96	18,325
R-squared: 0,949		

Source: Authors

Table 4.20 Results of
regressing Y against both $X1$
and $X2$

Variable	Coefficient	t-Statistic
Intercept	−0,38	−0,526
$X1$	5,05	83,455
$X2$	1,04	18,595
R-squared: 0,998		

Source: Authors

$X1$ equals 4,96 in the reduced regression as compared to 5,05 in the full regression—the difference of less than 2%.

4.5.4 ARCH-LM test for the Heteroscedasticity of Residuals

This tool can be applied for testing the variance of residuals in time-series models. If the variance of regression residuals tends to change (rise or fall) with time, then squared values of residuals will also show trends. If this is the case, then there is a relationship between current unlagged squared values of residuals and lagged squared values of residuals. Therefore, we can apply the simple test for heteroscedasticity, based on multiple regression, in which squared values of our residuals constitute the dependent variable and lagged squared values of residuals constitute the set of explanatory variables. This test is very similar to F-test for autocorrelation.

We remember from the F-test for the statistical significance of the whole set of explanatory variables, that if at least one explanatory variable is statistically significant, then the F statistic points to the statistical significance of the whole regression. So, if the F statistic for such regression of squared residuals against lagged squared residuals indicates that this whole regression is statistically significant, then we receive the information that some heteroscedasticity of residuals exists. The procedure of *ARCH-LM* test for heteroscedasticity runs as follows (Engle 1982) (Table 4.21).

In checking the presence of residuals heteroscedasticity in our model by means of the *ARCH-LM* test, the first thing to do is to prepare the data by squaring and lagging our residuals to the specified maximum considered lag (k). In this problem we will consider the lags of up to fourth order. The data prepared for the regression of squared residuals against lagged squared residuals are shown below (Table 4.22).

Table 4.21 The procedure of *ARCH-LM* test for the heteroscedasticity of residuals

Step 1	Formulate the hypotheses: $H_0 : r_{e_t^2, e_{t-1}^2} = r_{e_t^2, e_{t-2}^2} = \ldots = r_{e_t^2, e_{t-k}^2} = 0$ (residuals are homoscedastic) $H_1 : r_{e_t^2, e_{t-1}^2} \neq 0 \lor r_{e_t^2, e_{t-2}^2} \neq 0 \lor \ldots \lor r_{e_t^2, e_{t-k}^2} \neq 0$ (residuals are heteroscedastic) where: e_t—vector of the regression residuals, $r_{e_t^2, e_{t-i}^2}$—correlation between squared residuals and squared residuals lagged by i-th order, k—maximum assumed lag of squared residuals.
Step 2	Regress the squared residuals against the lagged squared residuals: $e_t^2 = \alpha_0 + \alpha_1 e_{t-1}^2 + \alpha_2 e_{t-2}^2 + \ldots + \alpha_k e_{t-k}^2$
Step 3	For the above regression of squared residuals against lagged squared residuals, compute the empirical statistic F according to the following formula: $F = \frac{R^2}{1-R^2} \frac{n-(k+1)}{k}$, where: R^2—coefficient of determination (R-squared) of the regression of squared residuals against lagged squared residuals, n—number of observations on which the regression of squared residuals was estimated, k—number of explanatory variables in the regression.
Step 4	Obtain the critical value F^* from the Fisher–Snedecor table, for $m_1 = k$, $m_2 = l - (k+1)$ and for assumed significance level α.
Step 5	Compare the values of the empirical F with critical F^*. If $F \leq F^*$, then H_0 is not rejected, which means that the regression' residuals cannot be concluded to be heteroscedastic. If $F > F^*$, then H_0 is rejected, and the data show that that there is some heteroscedasticity of the regression' residuals.

Source: Authors

The regression of the squared residuals against the lagged squared residuals has the following form:

$$e_t^2 = \alpha_0 + \alpha_1 e_{t-1}^2 + \alpha_2 e_{t-2}^2 + \alpha_3 e_{t-3}^2 + \alpha_4 e_{t-4}^2.$$

The results of this regression are presented in Table 4.23.

Although our model was estimated on the basis of 20 observations, in testing the heteroscedasticity by *ARCH-LM* test we have only 16 observations, since four observations were lost by lagging squared residuals. This means that for our *ARCH-LM* test for heteroscedasticity $n = 16$, there are four explanatory variables (which means that $k = 4$) and the coefficient of determination $R^2 = 0,379$. Putting these numbers into the formula for the empirical statistic F produces the following results:

$$F = \frac{R^2}{1 - R^2} \frac{n - (k+1)}{k} = \frac{0,379}{1 - 0,379} \frac{16 - (4+1)}{4} = 1,67.$$

Now we obtain the critical value of F^*. For $n = 16$ and $k = 4$,

Table 4.22 The residuals, squared residuals, and lagged squared residuals from our model

Residuals e_t	Squared residuals e_t^2	First lag of squared residuals e_{t-1}^2	Second lag of squared residuals e_{t-2}^2	Third lag of squared residuals e_{t-3}^2	Fourth lag of squared residuals e_{t-4}^2
−178,78	31.962,27	–	–	–	–
9,80	96,12	31.962,27	–	–	–
50,33	2.532,61	96,12	31.962,27	–	–
36,99	1.367,97	2.532,61	96,12	31.962,27	–
0,75	0,57	1.367,97	2.532,61	96,12	31.962,27
−187,76	35.253,34	0,57	1.367,97	2.532,61	96,12
48,92	2.392,86	35.253,34	0,57	1.367,97	2.532,61
−47,09	2.217,71	2.392,86	35.253,34	0,57	1.367,97
−41,41	1.714,81	2.217,71	2.392,86	35.253,34	0,57
244,71	59.884,85	1.714,81	2.217,71	2.392,86	35.253,34
25,48	649,42	59.884,85	1.714,81	2.217,71	2.392,86
250,72	62.860,90	649,42	59.884,85	1.714,81	2.217,71
96,62	9.335,83	62.860,90	649,42	59.884,85	1.714,81
−320,80	102.910,94	9.335,83	62.860,90	649,42	59.884,85
−41,51	1.722,80	102.910,94	9.335,83	62.860,90	649,42
−216,89	47.042,50	1.722,80	102.910,94	9.335,83	62.860,90
−240,18	57.688,60	47.042,50	1.722,80	102.910,94	9.335,83
74,38	5.532,13	57.688,60	47.042,50	1.722,80	102.910,94
150,58	22.673,61	5.532,13	57.688,60	47.042,50	1.722,80
285,14	81.305,06	22.673,61	5.532,13	57.688,60	47.042,50

Source: Authors

Table 4.23 Results of regressing squared residuals against lagged (up to fourth order) squared residuals

Variable	Coefficient	t-Statistic
Intercept	21.210,58	1,449
First lag e_{t-1}^2	−0,54	−1,787
Second lag e_{t-2}^2	0,13	0,441
Third lag e_{t-3}^2	0,46	1,598
Fourth lag e_{t-4}^2	0,40	1,324
R-squared: 0,379		

Source: Authors

$$m_1 = k = 4,$$

$$m_2 = n - (k+1) = 16 - (4+1) = 11.$$

For the level of statistical significance set at $\alpha = 0,05$, and computed values of m_1 and m_2, the critical F^* from the Fisher–Snedecor table equals 3,36.

Finally, we need to compare the value of empirical statistic F with its critical value F^*. In our case $F < F^*$, which means that H_0 is not rejected, and we conclude that there is not any significant heteroscedasticity of the residuals in our model.

4.5.5 Breusch–Pagan Test for the Heteroscedasticity of Residuals

The Breusch–Pagan test is recommended for situations in which it is justified to suspect that the variance of the residuals may be correlated with the values of the explanatory variables. The test is more universal than *ARCH-LM* test because it can be applied to both time-series as well as cross-sectional models. The procedure of Breusch–Pagan test runs as follows (Breusch and Pagan 1979) (Table 4.24).

In checking for the presence of heteroscedasticity of residuals in our model by means of the Breusch–Pagan test, the first thing to do is to prepare the data by computing the values of r_n. These computations are shown in Table 4.25.

Then r_n is regressed against the set of explanatory variables from our model. The results of this regression are presented in Table 4.26.

The above regression of r_n against NS_t, PRM_t, and $NPOS_t$ was estimated on the basis of 20 observations. Thus, for the Breusch–Pagan test of our model $n = 20$, there are three explanatory variables (which means that $k = 3$) and the coefficient of determination is $R^2 = 0,254$. Putting these numbers into the formula for the empirical statistic F produces the following result:

$$F = \frac{R^2}{1 - R^2} \frac{n - (k + 1)}{k} = \frac{0,254}{1 - 0,254} \frac{20 - (3 + 1)}{3} = 1,82.$$

Now we need to obtain the critical value of F^*. For $n = 20$ and $k = 3$,

$$m_1 = k = 3,$$

$$m_2 = n - (k + 1) = 20 - (3 + 1) = 16.$$

For the level of statistical significance set at $\alpha = 0,05$, and computed values of m_1 and m_2, the critical value of F^* from the Fisher–Snedecor table equals 3,24.

Finally, we compare the value of empirical statistic F with its critical value F^*. In our case $F < F^*$, which means that H_0 is not rejected, and we conclude that there is not any significant heteroscedasticity of the residuals in our model.

Table 4.24 The procedure of the Breusch–Pagan test for the heteroscedasticity of residuals

Step 1	Formulate the hypotheses: $H_0 : r_{e^2,x_{1t}} = r_{e^2,x_{2t}} = \ldots = r_{e^2,x_{kt}} = 0$ (residuals are homoscedastic) $H_1 : r_{e^2,e^2_{t-1}} \neq 0 \vee r_{e^2,e^2_{t-2}} \neq 0 \vee \ldots \vee r_{e^2,e^2_{t-k}} \neq 0$ (residuals are heteroscedastic) where: e—vector of the regression residuals, $x_1, x_2, \ldots, x_k$—set of k explanatory variables of the tested model, $r_{e^2,x_{it}}$—correlation between squared residuals from the tested model and i-th explanatory variable.
Step 2	For each residual from the tested model, compute the value of r_n as follows: $r_n = \dfrac{e_n^2}{S^2} - 1,$ where: e_n—n-th residual from the tested model, S^2—variance of the residuals from the tested model.
Step 3	Regress r_n against the set of explanatory variables from the tested model.
Step 4	For the above regression of r_n against the set of explanatory variables from the tested model, compute the empirical statistic F according to the following formula: $F = \dfrac{R^2}{1-R^2} \dfrac{n-(k+1)}{k},$ where: R^2—coefficient of determination (R-squared) of regression of r_n against the set of explanatory variables from the tested model, n—number of observations on which the regression of r_n was estimated (the same as in the tested model), k—number of explanatory variables in the regression of r_n (the same as in the tested model).
Step 5	Obtain the critical (theoretical) value F^* from the Fisher–Snedecor table for $m_1 = k$, $m_2 = l - (k+1)$ and for the assumed significance level α.
Step 6	Compare the values of the empirical F and the critical F^*. If $F \leq F^*$, then H_0 is not rejected, which means that there is not enough evidence to conclude that the variance of residuals in the tested model is heteroscedastic. If $F > F^*$, then H_0 is rejected, which means that the variance of residuals in the tested model is heteroscedastic.

Source: Authors

4.6 Testing the Symmetry of Regression Residuals

4.6.1 Illustration of Non-symmetry of Regression Residuals

Again, we illustrate the problem of non-symmetry of residuals with a numerical example. Consider the series of residuals from a time-series model with 20 observations. For this series of residuals, we again run the four tests discussed above:

- Test for outlying residuals based on two-sigma range.
- F-test for autocorrelation.
- ARCH-LM test for heteroscedasticity.
- Jarque–Bera (JB) test for normality of distribution.

Table 4.25 The computations of r_n for Breusch–Pagan test (for heteroscedasticity of residuals)

Residuals e_t	Squared residuals e_t^2	$r_n = \frac{e_t^2}{S^2} - 1$
−178,78	31.962,27	0,1477
9,80	96,12	−0,9965
50,33	2.532,61	−0,9091
36,99	1.367,97	−0,9509
0,75	0,57	−1,0000
−187,76	35.253,34	0,2658
48,92	2.392,86	−0,9141
−47,09	2.217,71	−0,9204
−41,41	1.714,81	−0,9384
244,71	59.884,85	1,1503
25,48	649,42	−0,9767
250,72	62.860,90	1,2571
96,62	9.335,83	−0,6648
−320,80	102.910,94	2,6952
−41,51	1.722,80	−0,9381
−216,89	47.042,50	0,6892
−240,18	57.688,60	1,0714
74,38	5.532,13	−0,8014
150,58	22.673,61	−0,1859
285,14	81.305,06	1,9194
Variance of residuals: $S^2 = 27.849,73$		

Source: Authors

Table 4.26 Results of regressing r_n against the explanatory variables from the tested model

Variable	Coefficient	t-Statistic
Intercept	−1,44	−0,165
NS$_t$	0,00	0,543
PRM$_t$	0,01	0,086
NPOS$_t$	−0,24	−1,677
R-squared: 0,254		

Source: Authors

Table 4.27 presents the investigated residuals together with their lags of up to fourth order.

We first check if there are any outliers. Table 4.28 presents the statistics based on the three-sigma rule.

As the table shows, there are no residuals with values outside the range determined by the two-sigma range. We can conclude, therefore, that the tested sample of residuals is free from any outliers.

Table 4.27 The tested residuals and lagged residuals up to the fourth order

Residuals e_t	First lag e_{t-1}	Second lag e_{t-2}	Third lag e_{t-3}	Fourth lag e_{t-4}
−2	−	−	−	−
−1	−2	−	−	−
−2	−1	−2	−	−
−1	−2	−1	−2	−
6	−1	−2	−1	−2
−2	6	−1	−2	−1
−1	−2	6	−1	−2
−4	−1	−2	6	−1
0	−4	−1	−2	6
−4	0	−4	−1	−2
6	−4	0	−4	−1
−1	6	−4	0	−4
0	−1	6	−4	0
−1	0	−1	6	−4
5	−1	0	−1	6
−2	5	−1	0	−1
−1	−2	5	−1	0
−1	−1	−2	5	−1
0	−1	−1	−2	5
6	0	−1	−1	−2

Source: Authors

We now test for the autocorrelation of residuals up to the fourth order. We first regress the residuals against the lagged residuals. This regression has the following form:

$$e_t = \alpha_0 + \alpha_1 e_{t-1} + \alpha_2 e_{t-2} + \alpha_3 e_{t-3} + \alpha_4 e_{t-4}.$$

The results of this regression are presented in Table 4.29.

Given the above results we can conclude that up to the fourth order, there is no significant autocorrelation within our sample of residuals. The value of the empirical F statistic is smaller than its critical value; thus, the regression of the investigated residuals against lagged residuals is statistically insignificant at 5% significance level. We then check the potential heteroscedasticity of these residuals. Table 4.30 presents the squared values of the tested residuals, and Table 4.31 contains the results of regressing these squared residuals against lagged squared residuals.

Given the above results, we can conclude that there is no significant heteroscedasticity of residuals. The value of the empirical F statistic is smaller than its critical value. Thus, the whole regression of the squared residuals against lagged squared residuals is statistically insignificant at 5% significance level.

Next we test the normality of the distribution of residuals. The first thing to do is to compute the values of B_1 and B_2. However, before we do this, we need to

Table 4.28 Checking for outlying residuals on the basis of the three-sigma rule

Arithmetic average of residuals	0
Standard deviation of residuals	3,15
Upper bound[a]	6,30
Lower bound[b]	−6,30
Highest residual	6
Lowest residual	−4

Source: Authors
[a]Arithmetic average of residuals + two standard deviations of residuals
[b]Arithmetic average of residuals − two standard deviations of residuals

Table 4.29 Results of regressing residuals against lagged (up to fourth order) residuals

Variable	Coefficient	t-Statistic
Intercept	0,14	0,170
First lag e_{t-1}	−0,50	−1,534
Second lag e_{t-2}	−0,30	−0,969
Third lag e_{t-3}	−0,56	−1,790
Fourth lag e_{t-4}	−0,16	−0,498

R-squared: 0,321
F statistic: 1,30
Critical F statistic (at 5% significance level): 3,36

Source: Authors

calculate the values of M_3, M_4, S, and A. The results of all these computations are shown in Table 4.32.

The following computations look as follows. For $n = 20$:

$$S = \sqrt{\frac{1}{n}\sum_{i=1}^{n} e_i^2} = \sqrt{\frac{188}{20}} \approx 3,06594,$$

$$M_3 = \frac{1}{n}\sum_{i=1}^{n} e_i^3 = \frac{606}{20} \approx 30,30,$$

$$M_4 = \frac{1}{n}\sum_{i=1}^{n} e_i^4 = \frac{5\,096}{20} \approx 254,80,$$

$$A = \frac{M_3}{S^3} = \frac{30,30}{(3,06594)^3} \approx 1,05136,$$

Table 4.30 The squared residuals and lagged squared residuals up to the fourth order

Residuals e_t^2	First lag e_{t-1}^2	Second lag e_{t-2}^2	Third lag e_{t-3}^2	Fourth lag e_{t-4}^2
4	–	–	–	–
1	4	–	–	–
4	1	4	–	–
1	4	1	4	–
36	1	4	1	4
4	36	1	4	1
1	4	36	1	4
16	1	4	36	1
0	16	1	4	36
16	0	16	1	4
36	16	0	16	1
1	36	16	0	16
0	1	36	16	0
1	0	1	36	16
25	1	0	1	36
4	25	1	0	1
1	4	25	1	0
1	1	4	25	1
0	1	1	4	25
36	0	1	1	4

Source: Authors

$$B_1 = A^2 = (1,05136)^2 \approx 1,1054,$$

$$B_2 = \frac{M_4}{S^4} = \frac{254,80}{(3,06594)^4} \approx 2,8837.$$

Now that we have computed the values of B_1 and B_2, we can easily obtain the value of the empirical Jarque–Bera statistic (JB_e) as follows:

$$JB_e = n\left(\frac{1}{6}B_1 + \frac{1}{24}(B_2 - 3)^2\right) = 20\left(\frac{1}{6}1,1054 + \frac{1}{24}(2,8837 - 3)^2\right) \approx 3,6958.$$

The next step is to derive the critical value of the *JB* statistic (JB^*), from the Chi-squared distribution table for 2 degrees of freedom and for the assumed level of statistical significance $\alpha = 5\%$.

$$JB^* = 5,99.$$

Given that in the case of our residuals $JB_e \leq JB^*$, we can conclude that these residuals are normally distributed.

To sum up, the investigated sample of residuals has passed all four tests:

Table 4.31 Results of regressing squared residuals against lagged squared residuals

Variable	Coefficient	t-Statistic
Intercept	31,74	4,140
First lag e_{t-1}^2	−0,53	−1,961
Second lag e_{t-2}^2	−0,75	−2,659
Third lag e_{t-3}^2	−0,49	−1,752
Fourth lag e_{t-4}^2	−0,48	−1,722

R-squared: 0,462
F statistic: 2,36
Critical F statistic (at 5% significance level): 3,36

Source: Authors

Table 4.32 The computations for Jarque–Bera test of normality of residuals' distribution

e_t	e_t^2	e_t^3	e_t^4
−2	4	−8	16
−1	1	−1	1
−2	4	−8	16
−1	1	−1	1
6	36	216	1.296
−2	4	−8	16
−1	1	−1	1
−4	16	−64	256
0	0	0	0
−4	16	−64	256
6	36	216	1.296
−1	1	−1	1
0	0	0	0
−1	1	−1	1
5	25	125	625
−2	4	−8	16
−1	1	−1	1
−1	1	−1	1
0	0	0	0
6	36	216	1.296
Σ	188	606	5.096

Source: Authors

- There are no outlying residuals.
- The residuals are not serially correlated (up to fourth order of autocorrelation).
- The variance of residuals seems to be homoscedastic (at least not correlated with time).
- The residuals are normally distributed.

However, let's look at the following bar chart of our sample of residuals (Fig. 4.15).

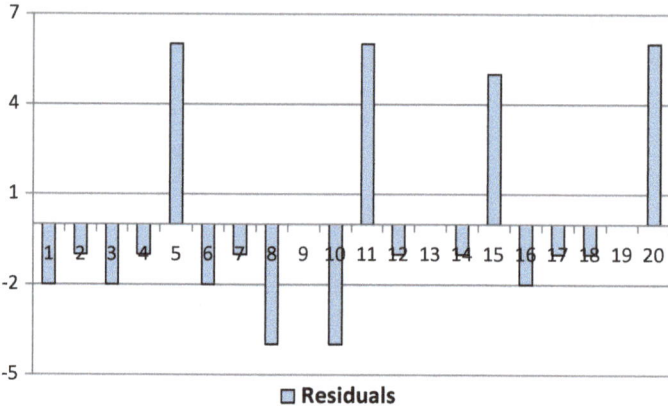

Fig. 4.15 The investigated residuals. Source: Authors

The visual inspection of the above figure suggests that the pattern of these residuals is not purely random. If residuals are to be chaotic (noisy), then something is wrong here. Namely, the number of negative residuals (13 out of 20, but of relatively small absolute values) is clearly overwhelming the number of positive residuals (only 4, but of relatively large absolute values).

If residuals are to be random, then the number of negative residuals should not differ significantly from the number of positive residuals. This condition is not satisfied here.

For ensuring robustness of the estimated regression, it is always recommended to test for the symmetry of residuals. We will show two simple procedures which can be applied here:

- Symmetry test.
- t-Student test of symmetry.

4.6.2   Symmetry Test

The procedure for the symmetry test runs as follows (Hellwig 1960) (Table 4.33).

In checking the symmetry of residuals in our model, the first thing to do is to compute the number of positive residuals (denoted as m_{e1}) and the number of negative residuals (denoted as m_{e2}). This is shown in Table 4.34.

Then we need to obtain the critical values of m_1 and m_2, from the table for symmetry test (included in the Appendix A5). For 20 residuals ($n = 20$) and for the assumed level of statistical significance α of 10%, we find that $m_1 = 6$ and $m_2 = 14$. Given that in our model $m_1 \leq m_{e1} \leq m_2$ as well as $m_1 \leq m_{e2} \leq m_2$, H_0 is not rejected, which means that the model's residuals are symmetrically distributed.

Table 4.33 The procedure for the symmetry test of residuals

Step 1	Formulate the hypotheses: $H_0 : \left[\frac{m}{n} = \frac{1}{2}\right]$ (residuals are symmetrically distributed) $H_1 : \left[\frac{m}{n} \neq \frac{1}{2}\right]$ (residuals are not symmetrically distributed) where: m—the number of residuals of the same sign (positive or negative), n—total number of residuals.
Step 2	Determine the number of positive residuals (denoted as m_{e1}) and the number of negative residuals (denoted as m_{e2}), out of total n residuals (neglect all residuals with value of zero).
Step 3	Derive the critical values of m_1 and m_2, from the table for symmetry test, for n residuals and for assumed significance level α (usually assumed at 10%).
Step 4	Compare m_{e1} and m_{e2} with m_1 and m_2. If both $m_1 \leq m_{e1} \leq m_2$ and $m_1 \leq m_{e2} \leq m_2$, then H_0 is not rejected, and there is no evidence against the random distribution of residuals. If either $m_{e1} < m_1$ or $m_{e2} < m_1$ or $m_{e1} > m_2$ or $m_{e2} > m_2$, then H_0 is rejected, which means that the regression residuals are not symmetrically distributed.

Source: Authors

Table 4.34 The computations for the symmetry test (for symmetry of residuals)

Residuals $(e_t = COGS_t - \widehat{COGS}_t)$	Sign of residual (1—positive, 0—otherwise)	Sign of residual (1—negative, 0—otherwise)
−178,78	0	1
9,80	1	0
50,33	1	0
36,99	1	0
0,75	1	0
−187,76	0	1
48,92	1	0
−47,09	0	1
−41,41	0	1
244,71	1	0
25,48	1	0
250,72	1	0
96,62	1	0
−320,80	0	1
−41,51	0	1
−216,89	0	1
−240,18	0	1
74,38	1	0
150,58	1	0
285,14	1	0
$m_{e1}=$	12	–
$m_{e2}=$	–	8

Source: Authors

Table 4.35 The procedure of t-Student test for symmetry of residuals

Step 1	Formulate the hypotheses: $H_0 : \left[\frac{m}{n} = \frac{1}{2}\right]$ (residuals are symmetrically distributed) $H_1 : \left[\frac{m}{n} \neq \frac{1}{2}\right]$ (residuals are not symmetrically distributed) where: m—the number of residuals of the same sign (positive or negative), n—total number of residuals.
Step 2	Determine the number of positive residuals (denoted as m_{e1}) and the number of negative residuals (denoted as m_{e2}), out of total n residuals (ignore all residuals with value of zero).
Step 3	Compute two empirical statistics according to the following formulas: $$t_{e1} = \frac{\left|\frac{m_{e1}}{n} - \frac{1}{2}\right|}{\sqrt{\frac{\frac{m_{e1}}{n}\left(1 - \frac{m_{e1}}{n}\right)}{n-1}}}$$ $$t_{e2} = \frac{\left|\frac{m_{e2}}{n} - \frac{1}{2}\right|}{\sqrt{\frac{\frac{m_{e2}}{n}\left(1 - \frac{m_{e2}}{n}\right)}{n-1}}}$$
Step 4	Obtain the critical value of t_t from the t-Student distribution table for $n-1$ degrees of freedom and for the assumed level of significance α.
Step 5	Compare t_{e1} with t_t and t_{e2} with t_t. If both $t_{e1} < t_t$ and $t_{e2} < t_t$, then H_0 is not rejected, which means that the residuals are symmetrically distributed. If either $t_{e1} \geq t_t$ or $t_{e2} \geq t_t$, then H_0 is rejected (and H_1 is true), and we conclude that the regression' residuals are not symmetrically distributed.

Source: Authors

4.6.3 t-Student Test of Symmetry

Another simple tool for verifying the symmetry of residuals is a test based on the t-Student distribution. The procedure of this test runs as follows (Hellwig 1960) (Table 4.35).

Let's check the symmetry of residuals in our model, by means of the t-Student test for symmetry. The first thing to do is to compute the number of positive residuals (denoted as m_{e1}) and the number of negative residuals (denoted as m_{e2}). From Table 4.34, we recall that $m_{e1} = 12$ and $m_{e2} = 8$.

Then we calculate the empirical statistics, which are as follows:

$$t_{e1} = \frac{\left|\dfrac{m_{e1}}{n} - \dfrac{1}{2}\right|}{\sqrt{\dfrac{\dfrac{m_{e1}}{n}\left(1 - \dfrac{m_{e1}}{n}\right)}{n-1}}} = \frac{\left|\dfrac{12}{20} - \dfrac{1}{2}\right|}{\sqrt{\dfrac{\dfrac{12}{20}\left(1 - \dfrac{12}{20}\right)}{20-1}}} = 0,890,$$

$$t_{e2} = \frac{\left| \dfrac{m_{e1}}{n} - \dfrac{1}{2} \right|}{\sqrt{\dfrac{\dfrac{m_{e1}}{n}\left(1 - \dfrac{m_{e1}}{n}\right)}{n-1}}} = \frac{\left| \dfrac{8}{20} - \dfrac{1}{2} \right|}{\sqrt{\dfrac{\dfrac{8}{20}\left(1 - \dfrac{8}{20}\right)}{20 - 1}}} = 0,890.$$

The next step is to obtain the critical value of t_t, from the t-Student distribution table for $n - 1 = 19$ degrees of freedom and for the assumed level of statistical significance $\alpha = 5\%$. This value of t_t equals 2,093. Given that in our model $t_{e1} < t_t$ as well as $t_{e2} < t_t$, H_0 is not rejected, which means that the model's residuals are symmetrically distributed.

4.7 Testing the Randomness of Regression Residuals

4.7.1 Illustration of Nonrandomness of Regression Residuals

We illustrate the problem of lacking randomness of residuals with a numerical example. Consider the series of residuals from a time-series model with 20 observations. For this series of residuals, we will run four tests:

- Test for outlying residuals based on two-sigma range.
- F-test for autocorrelation.
- ARCH-LM test for heteroscedasticity.
- Jarque–Bera (JB) test for normality of distribution.

Table 4.36 presents the investigated residuals together with their lags of up to fourth order.

First, check if in our sample of residuals there are any outliers. Table 4.37 presents the statistics based on two-sigma range.

As the above table shows, there are no residuals in this sample which have values outside the two-sigma range. We conclude, therefore, that the tested sample of residuals is free from any outliers.

We now test for the autocorrelation of up to fourth order. We regress the residuals against the lagged residuals. This regression has the following form:

$$e_t = \alpha_0 + \alpha_1 e_{t-1} + \alpha_2 e_{t-2} + \alpha_3 e_{t-3} + \alpha_4 e_{t-4}.$$

The results of this regression are presented in Table 4.38.

Given the above results, we conclude that there is no significant autocorrelation (of up to fourth order) within our sample of residuals. The empirical F statistic is smaller than its critical value, and we conclude that the regression of the investigated residuals against lagged residuals is statistically insignificant at the 5% significance level. We now check the potential heteroscedasticity of these residuals. Table 4.39 presents the squared values of the tested residuals, and Table 4.40

Table 4.36 The tested residuals and lagged residuals (up to the fourth order)

Residuals e_t	First lag e_{t-1}	Second lag e_{t-2}	Third lag e_{t-3}	Fourth lag e_{t-4}
−2	−	−	−	−
1	−2	−	−	−
4	1	−2	−	−
1	4	1	−2	−
4	1	4	1	−2
1	4	1	4	1
2	1	4	1	4
6	2	1	4	1
1	6	2	1	4
1	1	6	2	1
−3	1	1	6	2
−1	−3	1	1	6
−5	−1	−3	1	1
−1	−5	−1	−3	1
−1	−1	−5	−1	−3
−4	−1	−1	−5	−1
−1	−4	−1	−1	−5
−6	−1	−4	−1	−1
−1	−6	−1	−4	−1
4	−1	−6	−1	−4

Source: Authors

contains the results of regressing these squared residuals against lagged squared residuals.

Given the above results, we conclude that there is no significant heteroscedasticity of residuals. The empirical F statistic is smaller than its critical value which means that the whole regression of the squared residuals against lagged squared residuals is statistically insignificant at the 5% significance level.

We now test for the normality of the residuals' distribution. The first thing to do is to compute the values of B_1 and B_2. However, before we do this, we need to calculate the values of M_3, M_4, S, and A. The results of all these computations are shown in Table 4.41.

The following computations look as follows. For $n = 20$:

$$S = \sqrt{\frac{1}{n}\sum_{i=1}^{n} e_i^2} = \sqrt{\frac{188}{20}} \approx 3,06594,$$

$$M_3 = \frac{1}{n}\sum_{i=1}^{n} e_i^3 = \frac{-24}{20} \approx -1,20,$$

Table 4.37 Checking for outlying residuals on the basis of the two-sigma range

Arithmetic average of residuals	0
Standard deviation of residuals	3,15
Upper bound[a]	6,30
Lower bound[b]	−6,30
Highest residual	6
Lowest residual	−6

Source: Authors
[a]Arithmetic average of residuals + two standard deviations of residuals
[b]Arithmetic average of residuals − two standard deviations of residuals

Table 4.38 Results of regressing residuals against lagged (up to fourth order) residuals

Variable	Coefficient	t-Statistic
Intercept	−0,05	−0,056
First lag e_{t-1}	0,22	0,620
Second lag e_{t-2}	0,42	1,317
Third lag e_{t-3}	0,16	0,400
Fourth lag e_{t-4}	−0,40	−1,170

R-squared: 0,278
F statistic: 1,06
Critical F statistic (at 5% significance level): 3,36

Source: Authors

$$M_4 = \frac{1}{n} \sum_{i=1}^{n} e_i^4 = \frac{4.364}{20} \approx 218,20,$$

$$A = \frac{M_3}{S^3} = \frac{-1,20}{(3,06594)^3} \approx -0,04164,$$

$$B_1 = A^2 = (-0,04164)^2 \approx 0,0017,$$

$$B_2 = \frac{M_4}{S^4} = \frac{218,20}{(3,06594)^4} \approx 2,4694.$$

Now that we have computed the values of B_1 and B_2, we can easily obtain the value of empirical Jarque–Bera statistic (JB_e) as follows:

$$JB_e = n\left(\frac{1}{6}B_1 + \frac{1}{24}(B_2 - 3)^2\right) = 20\left(\frac{1}{6}0,0017 + \frac{1}{24}(2,4694 - 3)^2\right) \approx 0,2404.$$

Table 4.39 The squared residuals and lagged squared residuals (up to the fourth order)

Residuals e_t^2	First lag e_{t-1}^2	Second lag e_{t-2}^2	Third lag e_{t-3}^2	Fourth lag e_{t-4}^2
4	–	–	–	–
1	4	–	–	–
16	1	4	–	–
1	16	1	4	–
16	1	16	1	4
1	16	1	16	1
4	1	16	1	16
36	4	1	16	1
1	36	4	1	16
1	1	36	4	1
9	1	1	36	4
1	9	1	1	36
25	1	9	1	1
1	25	1	9	1
1	1	25	1	9
16	1	1	25	1
1	16	1	1	25
36	1	16	1	1
1	36	1	16	1
16	1	36	1	16

Source: Authors

Table 4.40 Results of regressing squared residuals against lagged squared residuals

Variable	Coefficient	t-Statistic
Intercept	31,71	3,680
First lag e_{t-1}^2	−0,68	−2,687
Second lag e_{t-2}^2	−0,52	−1,694
Third lag e_{t-3}^2	−0,51	−1,457
Fourth lag e_{t-4}^2	−0,63	−2,131

R-squared: 0,499
F statistic: 2,73
Critical F statistic (at 5% significance level): 3,36

Source: Authors

The next step is to obtain the critical value of JB statistic (JB^*), from the Chi-squared distribution table, for 2 degrees of freedom and for the assumed level of statistical significance α, which we assume at 5%.

$$JB^* = 5,99.$$

Table 4.41 The computations for Jarque–Bera test of normality of residuals' distribution

e_t	e_t^2	e_t^3	e_t^4
−2	4	−8	16
1	1	1	1
4	16	64	256
1	1	1	1
4	16	64	256
1	1	1	1
2	4	8	16
6	36	216	1.296
1	1	1	1
1	1	1	1
−3	9	−27	81
−1	1	−1	1
−5	25	−125	625
−1	1	−1	1
−1	1	−1	1
−4	16	−64	256
−1	1	−1	1
−6	36	−216	1.296
−1	1	−1	1
4	16	64	256
Σ	188	−24	4.364

Source: Authors

Given that in the case of our residuals $JB_e \le JB^*$, we conclude that the residuals do not depart significantly from the normal distribution.

To sum up, the investigated sample of residuals has passed all four tests, which means that:

- There are no outlying residuals.
- The residuals are not serially correlated (up to fourth order of autocorrelation).
- The variance of residuals seems to be homoscedastic (at least not correlated with time).
- The residuals are normally distributed.

However, the visual inspection of the chart below suggests that the pattern of these residuals is not purely random (Fig. 4.16).

If residuals are to be chaotic (noisy), then evidently something is wrong here. Namely, the long series of only positive residuals (from the second to the tenth residual) is followed by an equally long series of only negative residuals (from the eleventh to the nineteenth residual).

Residuals should be random, similar to the results obtained when tossing a coin. The direction of a given residual should give no information about the probable direction of adjacent residuals. This condition is not satisfied here.

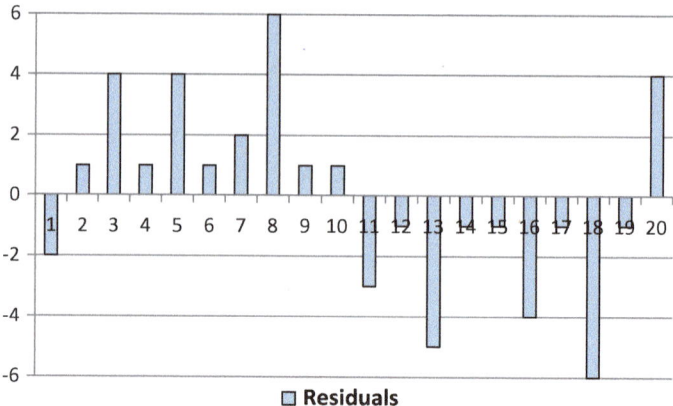

Fig. 4.16 The investigated residuals. Source: Authors

Table 4.42 The procedure of maximum series length test for randomness of residuals

Step 1	Formulate the hypotheses: H_0: residuals are randomly distributed H_1:residuals are not randomly distributed
Step 2	Denote positive residuals (as, e.g., "A") and negative residuals (as, e.g., "B"), neglecting all residuals with value of zero.
Step 3	Compute the length of the longest series of residuals of the same sign, denoted as k_e.
Step 4	Obtain the critical value of k_t from the maximum series length table, for n residuals (observations), and for the assumed level of significance α.
Step 5	Compare k_e with k_t. If $k_e < k_t$, then H_0 is not rejected, which means that the residuals are randomly distributed. If $k_e \geq k_t$, then H_0 is rejected and we conclude that the residuals are not randomly distributed.

Source: Authors

For ensuring the robustness of the estimated regression, it is recommended to test for the randomness of residuals. We will show two simple procedures which can be applied here:

- Maximum series length test.
- Number of series test.

4.7.2   Maximum Series Length Test

The procedure of maximum series length test runs as follows (Hellwig 1960) (Table 4.42).

Table 4.43 The computations for the maximum series length test (for randomness of residuals)

Residuals $(e_t = COGS_t - \widehat{COGS}_t)$	Sign of residual ("A"—positive, "B"—negative)	Length of the series of residuals of the same sign
−178,78	B	1
9,80	A	
50,33	A	
36,99	A	
0,75	A	4
−187,76	B	1
48,92	A	1
−47,09	B	
−41,41	B	2
244,71	A	
25,48	A	
250,72	A	
96,62	A	4
−320,80	B	
−41,51	B	
−216,89	B	
−240,18	B	4
74,38	A	
150,58	A	
285,14	A	3
Length of the longest series of residuals (k_e)		4

Source: Authors

Let's check the randomness of residuals in our model by means of the maximum series length test. Denote the positive residuals (as "A") and negative residuals (as "B"). Then we compute the length of the longest series of residuals of the same sign (k_e). This is shown in the following table (Table 4.43).

We then obtain the critical value of k_t from the maximum series length table (included in the Appendix A6). For 20 residuals ($n = 20$) and for the assumed level of statistical significance α of 5%, we find that $k_t = 7$. Given that in our model $k_e < k_t$, H_0 is not rejected, which means that the model's residuals are randomly distributed.

4.7.3 Number of Series Test

Another simple tool for verifying the randomness of residuals is the test based on the number of series. The procedure of this test runs as follows (Hellwig 1960) (Table 4.44).

Table 4.44 The procedure of number of series test for randomness of residuals

Step 1	Formulate the hypotheses: H_0: residuals are randomly distributed H_1:residuals are not randomly distributed
Step 2	Denote positive residuals (as, e.g., "A") and negative residuals (as, e.g., "B").
Step 3	Compute the number of series of residuals of the same sign, denoted as k_e.
Step 4	Obtain the critical values of k_1 and k_2, from the tables for number of series test, for n_1 residuals of one side and for n_2 residuals of the opposite side.
Step 5	Compare k_e with k_1 and k_2. If $k_1 \leq k_e \leq k_2$, then H_0 is not rejected, which means that the residuals are randomly distributed. If $k_e < k_1$ or $k_e > k_2$, then H_0 is rejected (and H_1 is true), which means that the regression residuals are not randomly distributed.

Source: Authors

Let's check the randomness of residuals in our model, by means of the number of series test. We first denote the positive residuals as "A" and negative residuals as "B." Then we compute the number of all series of residuals of the same sign (k_e). This is shown in Table 4.45.

We finally obtain the critical values of k_1 and k_2 from the table for the number of series test (Appendix A7). For 12 positive residuals ($n_1 = 12$) and for 8 negative residuals ($n_2 = 8$), we find that $k_1 = 6$ and $k_2 = 14$. Given that in our model $k_1 \leq k_e \leq k_2$, H_0 is not rejected, concluding that the model's residuals are randomly distributed.

4.8 Testing the Specification of the Model: Ramsey's *RESET* Test

The Ramsey's *RESET* test ("Regression Equation Specification Error Test") is used to verify the correctness of the model specification (Ramsey 1969). The tool is designed to signal model misspecification caused by the following problems:

- Omission of some significant explanatory variables.
- Application of linear functional form to nonlinear relationships.
- Incorrect lags of the dependent variable (in autoregressive models).

When this test suggests model misspecification, it does not inform about the specific problem encountered. It only suggests that one (or more) of the three above causes of misspecification has occurred. Therefore, if Ramsey's *RESET* test states that model is incorrectly specified, it is recommended to do the following:

- Consider adding one or more explanatory variables to the tested model.

Table 4.45 The computations for the number of series test (for randomness of residuals)

Residuals ($e_t = \text{COGS}_t - \widehat{\text{COGS}_t}$)	Sign of residual ("A"— positive, "B"—negative)	Change of the series of residuals of the same sign
−178,78	B	
9,80	A	1
50,33	A	
36,99	A	
0,75	A	
−187,76	B	1
48,92	A	1
−47,09	B	1
−41,41	B	
244,71	A	1
25,48	A	
250,72	A	
96,62	A	
−320,80	B	1
−41,51	B	
−216,89	B	
−240,18	B	
74,38	A	1
150,58	A	
285,14	A	
Number of series of residuals (k_e)		7

Source: Authors

- Change the functional form of the model from linear to some nonlinear (e.g., exponential, power, or logarithmic).
- Change the structure of autoregressive lags (in autoregressive models).

The Ramsey's *RESET* test is run in accordance to the following procedure (Table 4.46).

Before we apply the Ramsey's *RESET* test to our model, we will illustrate it with some fictitious data. These will be the same data as in the example of autocorrelation in Sect. 4.4, caused by incorrect functional form. In that example, the clearly nonlinear relationship was captured by the linear regression function, with the resulting autocorrelation of residuals. This relationship is presented again in the following figure.

To test the correctness of specification of this linear regression model, we compute the fitted values of its dependent variable. Then we add auxiliary explanatory variables constructed as the respective powers of the fitted values from the tested linear regression. Setting the number of such auxiliary variables is subjective; however, we will add two such variables:

Table 4.46 The procedure of Ramsey's *RESET* test for the model specification

Step 1	Formulate the hypotheses: H_0: model is correctly specified H_1: model is misspecified
Step 2	Reestimate the tested regression (with all its dependent and explanatory variables) after adding the preset number of auxiliary explanatory variables. In this case, the observations are formed as the respective powers of fitted values of dependent variable from the tested regression. The number of such auxiliary variables is subjective, but usually two variables are added: squared fitted values from the tested model and cubic fitted values from the tested model.
Step 3	Compute the empirical statistic F according to the following formula: $$F = \frac{\left(R_2^2 - R_1^2\right)/2}{\left(1 - R_2^2\right)/(n - (k+3))},$$ where: R_2^2—coefficient of determination (R-squared) of the regression estimated in Step 2, R_1^2—coefficient of determination (R-squared) of the tested regression, n—number of observations on which the tested model was estimated, k—number of explanatory variables in the tested regression.
Step 4	Obtain the critical (theoretical) value F^* from the Fisher–Snedecor table, for $m_1 = 2$, $m_2 = n - (k+3)$ and for the assumed significance level α.
Step 5	Compare the empirical F with the critical F^*. If $F \leq F^*$, then H_0 is not rejected, which means that none of the auxiliary variables added in Step 2 is statistically significant. Thus, the conclusion that the tested model is correctly specified is supported. If $F > F^*$, then H_0 is rejected, which means that at least one of the auxiliary variables added in Step 2 is statistically significant, concluding that the tested model is misspecified.

Source: Authors

Table 4.47 Auxiliary regression estimated for Ramsey's *RESET* test conducted for the linear regression of Y against X

Variable	Coefficient	t-Statistic
Intercept	−8,73	−7,561
X	7,96	11,958
FITTED^2	0,01	8,130
FITTED^3	0,00	4,966
R-squared: 0,999		

Source: Authors

- Squared fitted values of the dependent variable obtained from the tested model (denoted as FITTED^2).
- Cubic fitted values of the dependent variable obtained from the tested model (denoted as FITTED^3).

Then we reestimate the tested linear regression after adding these two auxiliary explanatory variables. The results of this auxiliary regression are presented in Table 4.47. Table 4.48 in turn shows the values of the dependent and explanatory

Table 4.48 Actual data (of Y and X), fitted values of Y, and auxiliary explanatory variables for Ramsey's *RESET* test

Y	$X1$	Fitted Y^a	FITTED^2	FITTED^3
1,4	1,0	−18,60	345,91	−6.433,41
1,0	1,1	−15,73	247,37	−3.890,59
1,8	1,2	−12,86	165,31	−2.125,43
2,1	1,3	−9,99	99,73	−995,99
2,8	1,4	−7,12	50,64	−360,33
3,4	1,5	−4,25	18,02	−76,51
4,7	1,6	−1,37	1,89	−2,60
4,3	1,7	1,50	2,24	3,35
6,2	1,8	4,37	19,07	83,26
6,7	1,9	7,24	52,38	379,09
7,8	2,0	10,11	102,17	1.032,76
8,8	2,1	12,98	168,45	2.186,21
11,3	2,2	15,85	251,20	3.981,38
11,8	2,3	18,72	350,44	6.560,21
13,7	2,4	21,59	466,16	10.064,63
14,7	2,5	24,46	598,36	14.636,59
17,3	2,6	27,33	747,04	20.418,02
19,8	2,7	30,20	912,20	27.550,85
24,3	2,8	33,07	1.093,84	36.177,04
23,0	2,9	35,94	1.291,97	46.438,50
28,2	3,0	38,81	1.506,58	58.477,19
29,7	3,1	41,69	1.737,66	72.435,04
31,3	3,2	44,56	1.985,23	88.453,98
35,8	3,3	47,43	2.249,28	106.675,95
40,3	3,4	50,30	2.529,82	127.242,90
39,8	3,5	53,17	2.826,83	150.296,76
45,2	3,6	56,04	3.140,33	175.979,46
51,7	3,7	58,91	3.470,30	204.432,95
53,3	3,8	61,78	3.816,76	235.799,15
55,8	3,9	64,65	4.179,70	270.220,02
60,3	4,0	67,52	4.559,12	307.837,49
69,8	4,1	70,39	4.955,02	348.793,49
75,2	4,2	73,26	5.367,41	393.229,96
79,7	4,3	76,13	5.796,27	441.288,84
83,3	4,4	79,00	6.241,62	493.112,07
89,8	4,5	81,87	6.703,44	548.841,58
95,3	4,6	84,75	7.181,75	608.619,32
99,8	4,7	87,62	7.676,54	672.587,22
105,2	4,8	90,49	8.187,82	740.887,21
114,7	4,9	93,36	8.715,57	813.661,24
120,3	5,0	96,23	9.259,80	891.051,24

Source: Authors

^{a}From the following regression: $Y = -47,31 + 28,71\,X$

Fig. 4.17 The actual
(nonlinear) relationship
between Y and X and linear
regression of Y against X.
Source: Authors

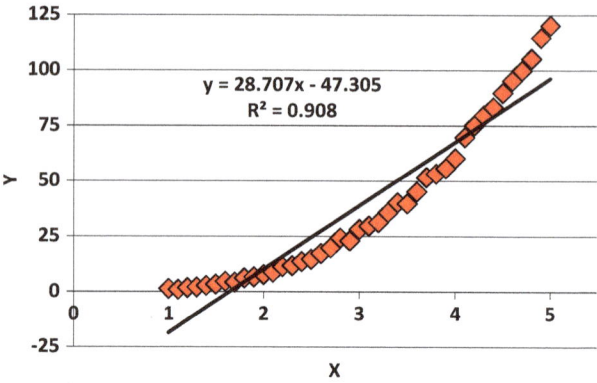

variable from the tested regression as well as the values of the two added auxiliary
explanatory variables.

Figure 4.17 shows that the R-squared of the tested model equals 0,908.
Table 4.47, in turn, informs us that the R-squared of the auxiliary regression equals
0,999. The tested regression of Y against X was estimated on the basis of 41 observations ($n = 41$), and it includes one explanatory variable (which means that $k = 1$).
Therefore, our empirical F statistic is computed as follows:

$$F = \frac{\left(R_2^2 - R_1^2\right)/2}{\left(1 - R_2^2\right)/(n - (k + 3))} = \frac{(0,999 - 0,908)/2}{(1 - 0,999)/(41 - (1 + 3))} = 1.377,76.$$

Now derive the critical value of F^*. For $n = 41$ and $k = 1$, we obtain:

$$m_1 = 2,$$

$$m_2 = n - (k + 3) = 41 - (1 + 3) = 37.$$

For the level of statistical significance set at $\alpha = 0,05$, and determined values of
m_1 and m_2, we can obtain the critical value of F^* from the Fisher–Snedecor table,
which equals 3,25.

Finally, we compare the value of the empirical statistic F with its critical value F^*.
In our case $F > F^*$, which means that H_0 is rejected, and we conclude that at least one
of the auxiliary variables is statistically significant. Thus, the tested model is
misspecified. Therefore, we can see that in this example, the Ramsey's *RESET* test
correctly suggests the misspecification of the tested regression, which was estimated
as linear for clearly nonlinear relationship.

Let's now verify the correctness of specification of our model, by means of the
Ramsey's *RESET* test. The first thing to do is to prepare the data by computing the
fitted values of the dependent variable as well as the values of auxiliary explanatory
variables. We will again add two auxiliary variables:

Table 4.49 The data for Ramsey's *RESET* test for correctness of specification of our model

$COGS_t$	NS_t	PRM_t	$NPOS_t$	Fitted $COGS_t$	$FITTED^2$	$FITTED^3$
8.700	11.200	100	2	8.878,78	78.832.733	699.938.489.263
10.525	13.400	105	2	10.515,20	110.569.349	1.162.658.387.791
10.235	12.700	107	3	10.184,67	103.727.603	1.056.431.917.304
11.750	15.000	110	3	11.713,01	137.194.694	1.606.963.362.898
11.145	14.500	109	5	11.144,25	124.194.218	1.384.050.905.122
12.235	16.900	107	4	12.422,76	154.324.935	1.917.141.430.166
11.925	16.400	105	5	11.876,08	141.041.352	1.675.018.825.113
11.420	16.000	104	6	11.467,09	131.494.211	1.507.856.288.238
12.530	17.500	106	5	12.571,41	158.040.355	1.986.790.131.083
12.240	15.800	108	3	11.995,29	143.886.891	1.725.964.436.677
14.155	19.000	111	2	14.129,52	199.643.228	2.820.862.215.919
15.165	20.300	113	3	14.914,28	222.435.725	3.317.468.523.452
14.275	18.500	115	2	14.178,38	201.026.400	2.850.228.275.857
14.575	19.600	115	1	14.895,80	221.884.779	3.305.150.698.747
15.765	21.900	113	3	15.806,51	249.845.652	3.949.186.962.986
15.050	21.000	110	1	15.266,89	233.078.017	3.558.377.102.241
14.655	20.000	111	0	14.895,18	221.866.521	3.304.742.770.591
14.330	19.000	110	0	14.255,62	203.222.749	2.897.066.633.466
14.800	20.000	108	0	14.649,42	214.605.577	3.143.847.746.086
14.295	19.000	107	0	14.009,86	196.276.166	2.749.801.519.429

Source: Authors

- Squared fitted values of the dependent variable obtained from our model (denoted as $FITTED^2$).
- Cubic fitted values of the dependent variable obtained from our model (denoted as $FITTED^3$).

These data are presented in Table 4.49. Table 4.50, in turn, presents the results of auxiliary regression estimated for Ramsey's RESET test for our model (i.e., the results of regressing $COGS_t$ against NS_t, PRM_t, $NPOS_t$, $FITTED^2$, and $FITTED^3$).

We know from Chap. 3 that the R-squared of our model equals 0,993. Table 4.50, in turn, informs us that the R-squared of the auxiliary regression equals 0,994. Our tested model was estimated on the basis of 20 observations ($n = 20$) and it includes three explanatory variables (which means that $k = 3$). Therefore, our empirical F statistic is computed as follows:

$$F = \frac{\left(R_2^2 - R_1^2\right)/2}{\left(1 - R_2^2\right)/(n - (k + 3))} = \frac{(0,994 - 0,993)/2}{(1 - 0,994)/(20 - (3 + 3))} = 1,16.$$

Now we need to obtain the critical value of F^*. For $n = 20$ and $k = 3$, we obtain:

$$m_1 = 2,$$

Table 4.50 Auxiliary regression estimated for Ramsey's *RESET* test conducted for our model

Variable	Coefficient	t-Statistic
Intercept	−11.407,13	−0,388
NS$_t$	0,87	0,493
PRM$_t$	126,53	0,492
NPOS$_t$	−179,41	−0,513
FITTED^2	0,00	−0,102
FITTED^3	0,00	0,028
R-squared: 0,994		

Source: Authors

$$m_2 = n - (k + 3) = 20 - (3 + 3) = 14.$$

For the level of statistical significance set at $\alpha = 0,05$, and determined values of m_1 and m_2, we obtain the critical value of F^* from the Fisher–Snedecor table, which equals 3,74.

Finally, we compare the value of the empirical statistic F with its critical value F^*. In our case $F < F^*$, which means that H_0 is not rejected, and we conclude that none of the two auxiliary variables is statistically significant. Therefore, we can conclude that our model is correctly specified.

4.9 Testing the Multicollinearity of Explanatory Variables

One of the main practical problems related to application of statistical and econometric methods in economic research is multicollinearity of explanatory variables. Multicollinearity means that one or more explanatory variables are strongly statistically related to other explanatory variables (Maddala 1992).

Optimally, good multiple regression model should include explanatory variables which:

• Have statistically significant relationships with the dependent variable.
• Do not show any statistically significant relationships with each other.

If these two conditions are satisfied, then any single explanatory variable has significant incremental "value added" in explaining the dependent variable, and it is not including information that may be contained by other explanatory variables. In contrast, if explanatory variables are related to each other, then some of these variables contain information that is included in the other variables.

Specifically, multicollinearity means that at least one of the explanatory variables shows a statistically significant linear relationship with at least one of the remaining explanatory variables. As we will show in the following example, multicollinearity heavily understates the t-Statistics of the explanatory variables, causing a paradox: the estimated regression may exhibit a large R-squared, and be statistically significant given its F statistic, while all the explanatory variables are statistically insignificant, given their t-Statistics.

This is very important when applying the "general to specific modeling" procedure, because when starting with the general model it often turns out that all candidate explanatory variables are statistically insignificant, while after reducing the model the variables begin to show statistical significance. As we will also show in another example, multicollinearity can distort not only t-Statistics but also the values of regression coefficients, often bringing puzzling and counterintuitive signs of the individual parameters.

4.9.1 Illustration of the Distorting Impact of Multicollinerity on t-Statistics

Let's suppose that our intention is to build the regression model of Y against a set of some explanatory variables. We propose variables $X1, X2, X3, \ldots, X7$ as our seven candidate explanatory variables. We want to select the final set of explanatory variables by means of "general to specific modeling" procedure. Table 4.51 presents all the data of dependent variable as well as all seven candidate explanatory variables. Table 4.52 presents the results of general regression of Y against all seven candidate explanatory variables.

As we can clearly see in Table 4.52, we obtained some strange results: all seven explanatory variables are statistically insignificant while the whole general model is strongly significant. The model has a relatively large R-squared, and its empirical F statistic is much larger that its critical value. This is the most common result (paradox) brought about by multicollinearity of explanatory variables.

Table 4.53 presents the specific model obtained after applying the "general to specific modeling" procedure. We have removed the candidate explanatory variables in the following order: first $X3$ (on the grounds of its smallest absolute value of t-Statistic, 0,231) and then $X6$, $X5$, and $X4$. Finally, we obtained the specific model with three statistically significant explanatory variables: $X1, X2,$ and $X7$.

As we can see from Table 4.53, we ended up with three explanatory variables ($X1, X2,$ and $X7$). These three explanatory variables are statistically significant and have empirical t-Statistics exceeding their respective critical value. However, these very same variables (i.e., $X1, X2,$ and $X7$) were not statistically significant when the general model was estimated! This is the typical result of multicollinearity: the variables which are significant in a reduced model are often found to be insignificant in broader models. It seems illogical because we could expect that adding insignificant variables to the model should not reduce the significance of other variables.

Multicollinearity constitutes one of the main problems of many regression models estimated on the basis of economic data, particularly of time-series type. It also poses problems on "general to specific modeling," because starting from

Table 4.51 The dependent variable and seven candidate explanatory variables

Y	X1	X2	X3	X4	X5	X6	X7
21	12	6	3	4	3	4	3
22	10	5	5	6	5	6	6
20	10	6	4	5	4	3	4
23	10	7	6	6	8	7	7
23	11	6	5	5	7	9	6
21	10	5	4	4	6	5	5
23	11	6	6	5	5	7	8
23	10	5	5	6	6	9	7
20	10	4	3	4	5	6	5
23	11	4	6	8	4	8	5
22	11	6	5	7	5	7	3
23	11	5	6	6	6	8	4
23	11	6	7	7	5	7	6
22	10	5	6	6	4	6	7
22	10	4	5	7	5	7	8
21	10	5	4	6	6	6	6
22	11	4	5	5	5	7	5
22	11	5	6	6	4	6	6
21	9	4	5	7	5	5	7
20	11	5	4	5	4	4	3
21	10	4	3	4	6	7	4
20	9	3	2	3	7	5	6
21	11	4	3	4	2	8	3
20	10	3	5	5	4	3	5
20	12	2	6	5	3	5	2
21	11	3	7	2	6	7	7
21	10	3	6	7	5	4	6
20	10	2	5	6	4	3	5
21	11	2	6	5	6	5	4
15	9	1	4	8	7	6	3

Source: Authors

general model we often need to select from explanatory variables which all have t-Statistics pointing to their statistical insignificance. Furthermore, we can never be certain in the process of "general to specific modeling" whether on the way we do not remove some variables which are in fact statistically significant, but only have understated values of t-Statistics because of multicollinearity. Therefore, it is always necessary to control for potential multicollinearity of explanatory variables when building and interpreting econometric models.

Table 4.52 Results of the general regression of Y against all seven candidate explanatory variables

Variable	Coefficient	t-Statistic
Intercept	3,49	0,407
X1	1,14	1,405
X2	0,34	1,577
X3	0,07	0,231
X4	0,11	0,538
X5	0,05	0,247
X6	0,04	0,236
X7	0,55	2,038

Critical value of t-Statistic (at 5% significance level): 2,074
R-squared: 0,800
F statistic: 12,58
Critical value of F statistic (at 5% significance level): 2,46

Source: Authors

Table 4.53 Results of the specific regression

Variable	Coefficient	t-Statistic
Intercept	4,14	1,517
X1	1,19	4,702
X2	0,34	2,714
X7	0,62	5,485

Critical value of t-Statistic (at 5% significance level): 2,056
R-squared: 0,780
F statistic: 30,68
Critical value of F statistic (at 5% significance level): 2,98

Source: Authors

4.9.2   Testing for Multicollinearity by Means of Variance Inflation Factor

If multicollinearity exists, then there exists some linear relationship between two or more explanatory variables. Therefore, we can apply a simple test for multicollinearity based on regressing given explanatory variable against the remaining explanatory variables. We need to run this test separately for all explanatory variables. If we have k explanatory variables, we need to run k tests for multicollinearity. The procedure of this test runs as follows (Kutner et al. 2004) (Table 4.54).

Let's run this test for the general model presented in Table 4.52, on the basis of the data shown in Table 4.51. In that general model, we have seven explanatory variables; thus, we need to run seven separate tests for multicollinearity. First, we estimate seven linear regressions for the individual explanatory variables from the general model. Then, for all these seven auxiliary regressions, we calculate R-squared and VIF_i statistics. The results of such auxiliary regressions are presented in Table 4.55.

Table 4.54 The procedure of variance inflation factor test for multicollinearity of explanatory variables

Step 1	Formulate the hypotheses: $H_0 : b_1 = b_2 = \ldots = b_{k-1} = 0$ (all slope parameters $= 0$) $H_1 : b_1 \neq 0 \vee b_2 \neq 0 \vee \ldots \vee b_{k-1} \neq 0$ (at least one slope parameter $\neq 0$) where: $b_1, b_2, \ldots, b_{k-1}$—slope parameters in the auxiliary regression of x_i against all the remaining $k-1$ explanatory variables from the model which is tested for multicollinearity, x_i—i-th out of k explanatory variables in the model which is tested for multicollinearity, k—number of explanatory variables in the model which is tested for multicollinearity.
Step 2	Regress x_i against the remaining $k-1$ explanatory variables from the model which is tested for multicollinearity.
Step 3	For the auxiliary regression estimated above, compute variance inflation factor (denoted as VIF$_i$) according to the following formula: $\text{VIF}_i = \frac{1}{1-R_i^2}$, where: R_i^2—coefficient of determination (R-squared) of the regression of x_i against the remaining $k-1$ explanatory variables from the model which is tested for multicollinearity.
Step 4	Compare computed VIF$_i$ with its critical value VIF$^* = 5$ (as a "rule of thumb"). If VIF$_i \leq 5$, then H_0 is not rejected, which means that no significant multicollinearity between x_i and the remaining $k-1$ explanatory variables from the tested model was found. If $VIF_i > 5$, then H_0 is rejected, which means that there exists statistically significant linear relationship between x_i and at least one of the remaining $k-1$ explanatory variables from the tested model.

Source: Authors

Table 4.55 Results of variance inflation factor tests run for explanatory variables from the general model

Auxiliary regression	R-squared	VIF$_i$ statistic
$X1$ against $X2, X3, X4, X5, X6, X7$	0,941	16,82
$X2$ against $X1, X3, X4, X5, X6, X7$	0,776	4,47
$X3$ against $X1, X2, X4, X5, X6, X7$	0,853	6,81
$X4$ against $X1, X2, X3, X5, X6, X7$	0,713	3,49
$X5$ against $X1, X2, X3, X4, X6, X7$	0,723	3,61
$X6$ against $X1, X2, X3, X4, X5, X7$	0,732	3,73
$X7$ against $X1, X2, X3, X4, X5, X6$	0,882	8,50
Critical value of VIF$_i$ statistic (as a "rule of thumb"): 5,00		

Source: Authors

The results show that the linear regression of $X1$ against the remaining six variables has a large coefficient of determination (0,941) as well as a high variance inflation factor (equaling 16,82 and larger than the commonly applied subjective threshold of 5). This means that there is strong and statistically significant

Table 4.56 Results of variance inflation factor tests run for explanatory variables in our model

Auxiliary regression	R-squared	VIF_i statistic
NS_t against PRM_t and $NPOS_t$	0,526	2,11
PRM_t against NS_t and $NPOS_t$	0,493	1,97
$NPOS_t$ against NS_t and PRM_t	0,165	1,20
Critical value of VIF_i statistic (as a "rule of thumb"): 5,00		

Source: Authors

multicollinearity between $X1$ and the remaining explanatory variables of the general model. A similar situation, although of a weaker magnitude, is observed in the case of $X3$ (where $\text{VIF}_i = 6, 81$) and $X7$ (where $\text{VIF}_i = 8, 50$). These statistically significant multicollinearities were responsible for understatement of all t-Statistics of all seven explanatory variables in the general model.

We now test the potential multicollinearity in our model. There are three explanatory variables in our model: NS_t (quarterly net sales), PRM_t (quarterly wholesale price of raw materials per package), and $NPOS_t$ (the number of new points of sale opened in a given quarter). This means that we need to run three separate tests for multicollinearity. The results of the auxiliary regressions are presented in Table 4.56. As we can see, our model is free from multicollinearity of explanatory variables, because all three empirical VIF_i statistics have values below the critical threshold.

4.10 What to Do If the Model Is Not Correct?

So far in this chapter we have discussed the tests applicable for verifying the correctness of the constructed regression model. However, another important issue is: what to do when one (or more) of the verification tests points to some incorrectness of the model?

Depending of the type of the problem detected, the different corrective activities should follow. Sometimes the correction of the model is not possible and we need to proceed to the application of the model in its current form, with relevant awareness of its flaws and resulting caveats related to model interpretation and application. The detailed discussion of the procedures useful in correcting the econometric models reaches beyond the scope of this book. However, Table 4.57 presents a summary of basic procedures which are usually applied when specific problems with the model are detected. Some of the possible corrective actions are discussed in following chapters of this book.

4.11 Summary of Verification of Our Model

After running a series of verification tests of our model, we can conclude as follows:

Table 4.57 Basic procedures applied when model incorrectness is detected

Issue	What if H_0 is rejected?	What to do if H_0 is rejected?	What to do if You cannot modify the model?
General statistical significance of the regression	None of explanatory variables has statistically significant relationship with dependent variable	Propose a completely new set of candidate explanatory variables and build new model	–
Statistical significance of the individual variable	Tested explanatory variable does not show any significant relationship with dependent variable (but this is valid only if its t-Statistic is not distorted by autocorrelation or heteroscedasticity).	Check (by other tests) if t-Statistics are not distorted by autocorrelation or heteroscedasticity. If t-Statistics are credible, remove insignificant explanatory variable from the set of explanatory variables and reestimate the model.	Be aware that presence of statistically insignificant explanatory variables in the model may bias estimates of other model parameters (as well as t-Statistics).
Normality of residuals distribution	Residuals are not normally distributed, which brings the following effects: – Some statistics used in model verification (e.g., t-Statistics and F statistics) may be distorted. – If the model is to be used in forecasting, then non-normality of residuals distribution results in wider expected ranges of forecast errors (and may result in unknown expected distribution of forecast errors).	Check (by other tests) if non-normality of residuals distribution: – Is caused by outliers remaining in the sample (then control for their presence, e.g., by removing them or by adding some dummy variables to the model[a]). – Is caused by incorrect functional form of the model (then try alternative functional forms[a]). – Is caused by missing significant explanatory variables (then consider adding new explanatory variables or some autoregressive element[a]).	Be careful when interpreting t-statistics and F statistics (because their values may be seriously distorted). Be aware that forecasts obtained from the model may have unknown expected forecast errors.
Autocorrelation of residuals	Residuals are correlated, which brings the following effects: – Some statistics used in model verification (e.g., t-Statistics) may be distorted. – If model is to be used	Check (by other tests) if autocorrelation of residuals: – Is caused by incorrect functional form of the model (then try alternative functional forms[a]).	Be careful when interpreting t-Statistics (because their values may be distorted) or reestimate the model with method which is immune to

(continued)

Table 4.57 (continued)

Issue	What if H_0 is rejected?	What to do if H_0 is rejected?	What to do if You cannot modify the model?
	in forecasting, then autocorrelation may result in biased forecasts.	– Is caused by missing significant explanatory variables (then consider adding new explanatory variables or some autoregressive element[a]). – Is caused by incorrect (or lack of) lags of autoregressive element (then change the structure of lags of dependent variable[a]). – Is caused by structural breaks in modeled relationship (then try to capture them by some dummy variables[a]).	autocorrelation (e.g., weighted least squares). Be aware that forecasts obtained from the model may be biased.
Heteroscedasticity of residuals	Variance of residuals is not stable, which brings the following effects: – Some statistics used in model verification (e.g., t-Statistics) may be distorted. – if model is to be used in forecasting, then heteroscedasticity may result in changing expected range of forecast errors.	Check (by other tests) if heteroscedasticity of residuals: – Is caused by incorrect functional form of the model (then try alternative functional forms[a]). – Is caused by missing significant explanatory variables (then consider adding new explanatory variables). – Is caused by structural breaks in modeled relationship (then try to capture them by some dummy variables[a]).	Be careful when interpreting t-Statistics (because their values may be distorted) or reestimate the model with method which is immune to heteroscedasticity (e.g., weighted least squares). Be aware that forecasts from the model may have unknown distribution of forecast errors.
Symmetry of residuals	The number of residuals of one sign is significantly larger than number of residuals of opposite sign.	Check (by other tests) if non-symmetry of residuals: – Is caused by outliers remaining in the sample (then control for their presence, e.g., by removing them or by adding some dummy variables to	Be aware that forecasts obtained from the model may be biased.

Table 4.57 (continued)

Issue	What if H_0 is rejected?	What to do if H_0 is rejected?	What to do if You cannot modify the model?
		the model[a]). – Is caused by incorrect functional form of the model (then try alternative functional forms[a]).	
Randomness of residuals	Adjacent residuals tend to have the same signs (which means that they are not purely random).	Check (by other tests) if nonrandomness of residuals: – Is caused by incorrect functional form of the model (then try alternative functional forms[a]). – Is caused by missing significant explanatory variables (then consider adding new explanatory variables or some autoregressive element[a]). – Is caused by incorrect (or lack of) lags of autoregressive element (then change the structure of lags of dependent variable[a]). – Is caused by structural breaks in modeled relationship (then try to capture them by some dummy variables[a]).	Be aware that forecasts obtained from the model may be biased.
General specification of the model	Tested model is misspecified and should be changed.	Try to: – Change the functional form of the model from linear to some nonlinear (e.g., exponential, power, logarithmic, etc.)[a]. – Consider adding new explanatory variables or some autoregressive element[a]. – Change the structure of lags of dependent variable in autoregressive model[a].	Be aware that forecasts obtained from the model may be biased.

(continued)

Table 4.57 (continued)

Issue	What if H_0 is rejected?	What to do if H_0 is rejected?	What to do if You cannot modify the model?
Multicollinearity of explanatory variables	At least one of explanatory variables has statistically significant relationships with another explanatory variable	Check if removing one of correlated explanatory variables improves the model (by removing multicollinearity) without significant reduction of its general quality (i.e., its goodness of fit, lack of autocorrelation, etc.).	Be careful when interpreting t-Statistics (because their values may be distorted). Be aware that individual slope parameters may be strongly distorted and may have counterintuitive values. Interpret slope parameters jointly, instead of individually[a].

Source: Authors
[a]Issues discussed in more detail in other chapters

- Model is statistically significant, as confirmed by its F statistic, and includes individually significant explanatory variables.
- Model has normally distributed residuals (which is confirmed by both Hellwig test as well as Jarque–Bera test).
- Model has residuals which are not free from serial autocorrelation (which is detected by both F-test for autocorrelation as well as Box–Pearce test).
- Model has residuals with stable (homoscedastic) variance (which is confirmed by both *ARCH-LM* test as well as Breusch–Pagan test).
- Model has residuals which are symmetrically distributed around zero (which is confirmed by both symmetry test as well as t-Student test for symmetry).
- Model has residuals which are randomly distributed around zero (which is confirmed by both maximum series length test as well as number of series test).
- Model is correctly specified and its functional form seems to be correct (which is confirmed by Ramsey's *RESET* test).
- Model is free from multicollinearity of explanatory variables (which is confirmed by variance inflation factor test).

In summary, the only significant problem detected during the verification process is the presence of serial autocorrelation of residuals. In the case of our model, the autocorrelation might be caused by:

- Either missing significant explanatory variables.
- Or incorrect functional form.
- Or structural breaks of modeled relationship.
- Or mix of all these factors.

However, the Ramsey's *RESET* test as well as both tests of symmetry have not suggested any problems related to applied linear functional form. Therefore, it is justified to suppose that detected autocorrelation of residuals is not brought about by nonlinearity of the relationships between our dependent variable and explanatory variables. Hence, if we want to correct our model, we should consider:

- Adding some new explanatory variables (perhaps including some dummy variables capturing quarterly seasonality).
- Testing for potential structural breaks in the modeled relationships (e.g., by means of dummy variables).

Both of these adjustments will be discussed in Chap. 5.

References

Box GEP, Pearce DA (1970) Distribution of residual autocorrelations in autoregressive integrated moving average time series models. J Am Stat Assoc 65:1509–1526

Breusch TS, Pagan AR (1979) A simple test for heteroscedasticity and random coefficient variation. Econometrica 47:1287–1294

Charemza WW, Deadman DF (1992) New directions in econometric practice. General to specific modelling, cointegration and vector autoregression. Edward Elgar, Aldershot

Cochrane D, Orcutt GH (1949) Applications of least-squares regressions to relationships containing autocorrelated error terms. J Am Stat Assoc 44:32–61

Engle RF (1982) Autoregressive conditional heteroscedasticity with estimates of the variance of United Kingdom inflation. Econometrica 50:987–1008

Greene WH (1997) Econometric analysis. Prentice-Hall, New York

Harvey AC (1990) The econometric analysis of time series. Philip Allan, Hemel Hempstead

Hellwig Z (1960) Regresja liniowa i jej zastosowanie w ekonomii (Linear regression and its application in economics). PWG, Warsaw

Jarque CM, Bera AK (1987) A test for normality of observations and regression residuals. Int Stat Rev 55:163–172

Kutner MH, Nachtsheim CJ, Neter J (2004) Applied linear regression models. Irwin McGraw-Hill, Boston

Maddala GS (1992) Introduction to econometrics. Prentice-Hall, Englewood Cliffs

Ramsey JB (1969) Tests for specification errors in classical linear least squares regression analysis. J R Stat Soc B 31:350–371

Verbeek M (2000) A guide to modern econometrics. Wiley, Chichester

White H (1980) A Heteroskedasticity-consistent covariance estimator and a direct test for heteroskedasticity. Econometrica 48:817–838

Chapter 5
Common Adjustments to Multiple Regressions

5.1 Dealing with Qualitative Factors by Means of Dummy Variables

Many variables which constitute causal factors for other variables are of qualitative nature. Their qualitative nature means that they cannot be directly measured by numbers. However, if they influence the behavior of other variables, they must be dealt somehow in process of regression model building. We might imagine hundreds of such qualitative variables; however, examples of the most commonly used in economic analyses are:

- Variables used in marketing research (mostly in cross-sectional models), relating to peoples' profiles and impacting peoples' behavior as consumers:

  - Sex (male vs. female).
  - Marital status (married vs. non-married).
  - Having children vs. not having children.
  - Occupation ("white-collar" vs. "blue-collar" vs. other).
  - Education (higher vs. secondary level vs. primary level).
  - Place of living (big city vs. small town vs. rural).

- Variables used in modeling economic and financial results of companies (both in time-series as well as in cross-sectional models):

  - Company's industry according to its SIC (Standard Industrial Classification) code (belonging to a given industry vs. belonging to other industries).
  - Company's stock market listing status (listed on stock exchange vs. belonging to private shareholders).
  - Company's shareholding status (state-owned vs. owned by private shareholders).
  - Occurrence or not of one-off events such as employees' strike in a period for which economic results of a company are reported.

© Springer International Publishing AG 2018
J. Welc, P.J.R. Esquerdo, *Applied Regression Analysis for Business*,
https://doi.org/10.1007/978-3-319-71156-0_5

- Variables used in modeling macroeconomic processes (mostly in time-series models):

  - Occurrence or not of a recession in a period.
  - Occurrence or not of a one-off event impacting inflation rate, such as flood or drought.
  - Seasonal factors (first quarter of a year vs. second quarter vs. third quarter vs. fourth quarter).

Such variables cannot be directly measured and many of them cannot even be ranked. For example, we cannot quantify the impact of peoples' education on their buying behavior in a following way:

- Assign a value of zero to anyone who ended his or her education on primary level
- Assign a value of unity to anyone who ended his or her education on secondary level.
- Assign a value of two to anyone with higher education.

If we used such an artificial variable with three possible values (0 vs. 1 vs. 2) to capture education, we could automatically assume that the difference between having primary and secondary education is the same as the difference between having secondary and higher education. Meanwhile, a person's buying behavior might be related to whether the person has higher education while buying behavior may be indifferent to whether someone has primary-level vs. secondary-level education. Or it may happen that a given product is purchased mostly by people with secondary education while others tend to avoid it, regardless of whether they have primary or university-level education. Capturing such three-variants qualitative variable by one artificial variable with three possible values could result in obtaining false findings suggesting that the variable is statistically insignificant or at least its slope parameter could be misleading as to the true impact of education on buying behavior. Therefore, in such cases we need to capture the impact of qualitative variables in different way.

Normally, such qualitative factors are dealt with by introducing a proper number of binary dummy variables to the regression (Pindyck and Rubinfeld 1998). These dummy variables will have a value of unity for one variant and value of zero otherwise. However, in the regressions with intercept, if we have k variants of a given variable, we must consider no more than $k-1$ dummy variables (Johnston 1984). Without providing any mathematical proof we only request the reader to remember that if we try to introduce k dummy variables for the factor with k variants (e.g., two dummy variables for peoples' sex) to the regression with intercept, then we will not obtain any parameter estimates at all, due to perfect multicollinearity between these very dummy variables, which breaks down the numerical processes of estimating parameters.

Dummy variables for qualitative factors for models with intercept are constructed in the following way:

- For variables with two variants introduce one dummy variable, e.g.:
  - For person's sex assign zero to any man and unity to any woman (or vice versa).
  - For marital status assign zero to anyone who is married and unity to anyone who is not married
  - For one-off event in time-series models assign unity to the period when this event occurred and zero to all other observations.
- For variables with three variants introduce two dummy variables, e.g., for education:
  - A dummy variable for higher education having value of unity for anyone with higher education and value of zero for anyone else.
  - A dummy variable for secondary education (having value of unity for anyone with secondary education and value of zero for anyone else).
- For variables with four variants introduce three dummy variables, e.g., for classifying companies as belonging to one of four sectors: manufacturing, agriculture, financial services, nonfinancial services:
  - A dummy variable for manufacturing with a value of unity for any company belonging to the manufacturing sector and value of zero otherwise.
  - A dummy variable for agriculture with a value of unity for any agricultural company and value of zero otherwise.
  - A dummy variable for financial services with a value of unity for any company belonging to the financial services sector and value of zero otherwise).

We will illustrate the use of such dummy variables with the example of modeling the impact of customers' profiles on their purchases of some pharmaceutical product. Suppose that we have collected the following data about 25 users of a drug:

- Value of annual purchases of a drug by a given user in the last calendar year (in USD).
- Age (in years, rounded to the integer).
- Sex (male or female).
- Marital status (married or non-married).
- Place of living (big city, small town, or rural area).

The value of annual purchases of a drug and user age are the two quantitative variables in this dataset. All three remaining variables are of a qualitative nature. We are interested in capturing the impact of peoples' age, sex, marital status, and place of living (our candidate explanatory variables) on how much money they spend annually on purchases of this drug (our dependent variable).

Both qualitative variables are easily measured. However, we need to use some artificial dummy variables for capturing the effect of the remaining three qualitative variables. We introduce the following dummy variables:

- For user sex—one dummy variable, having value of zero for any man and value of unity for any woman in the sample.
- For user marital status—one dummy variable, having value of unity for anyone married and value of zero for anyone not married.
- For user place of living—two dummy variables:

  - Dummy variable, having value of unity for anyone living in a big city and value of zero otherwise.
  - Dummy variable, having value of unity for anyone living in a small town and value of zero otherwise.

These data are presented in Table 5.1.

Table 5.1 Selected personal data of 25 users of a drug

Number of persons	Value of annual purchases of a drug (in USD)	Age (in years)	Sex (1 if woman, 0 if man)	Marital status (1 if married, 0 if not married)	Living in a big city (1 if in a big city, 0 otherwise)	Living in a small town (1 if in a small town, 0 otherwise)
1	120	20	1	1	0	1
2	150	18	0	1	0	0
3	175	30	1	0	1	0
4	100	33	1	0	0	1
5	125	28	0	0	0	1
6	180	32	1	1	0	0
7	177	19	1	0	1	0
8	128	24	0	1	0	0
9	96	19	1	1	0	1
10	210	25	1	0	1	0
11	160	32	1	1	0	0
12	95	34	0	0	0	1
13	136	42	0	0	0	0
14	150	40	1	0	0	1
15	240	45	1	1	0	0
16	201	26	1	0	1	0
17	144	24	0	0	1	0
18	115	28	0	1	0	1
19	185	35	1	1	1	0
20	150	34	1	1	1	0
21	190	40	0	0	0	0
22	210	35	0	1	1	0
23	260	45	0	0	1	0
24	240	50	1	0	1	0
25	269	62	0	1	0	1

Source: Authors

We begin with a general model with all five candidate explanatory variables. The results of its estimation are presented in Table 5.2.

As can be seen in Table 5.2, the general regression of value of annual purchases of a drug against all five candidate explanatory variables is statistically significant but has some structural parameters which seem to be statistically insignificant. Therefore, we should reduce this general model by dropping explanatory variables which are not statistically significant. According to the "general to specific modeling" procedure, we excluded these variables in the following order: first Sex, then Marital status, and finally, Living in a big city. As a result, we obtained the specific model with two explanatory variables: Age and Living in a small town. The results of its estimation are shown in Table 5.3.

As can be seen in Table 5.3, the value of annual purchases of this drug is statistically related to two explanatory variables. The value of purchases tends to increase with persons' age. With every year of age the value of purchases rises by USD 3,01 annually, on average.

Table 5.2 Results of a general regression of value of annual purchases of a drug (in USD) against all five candidate explanatory variables (estimated on the basis of data from Table 5.1)

Variable	Coefficient	t-Statistic
Intercept	54,38	2,070
Age (in years)	3,13	5,262
Sex (1 if woman, 0 if man)	3,77	0,291
Marital status (1 if married, 0 if not married)	12,53	0,951
Living in a big city (1 if in a big city, 0 otherwise)	33,31	2,008
Living in a small town (1 if in a small town, 0 otherwise)	−32,08	−1,986
Critical value of t-Statistic (at 5% significance level): 2,093		
R-squared: 0,708		
F statistic: 19,08		
Critical value of F statistic (at 5% significance level): 2,74		

Source: Authors

Table 5.3 Results of a specific regression of value of annual purchases of a drug (in USD), estimated on the basis of data from Table 5.1

Variable	Coefficient	t-Statistic
Intercept	86,02	4,021
Age (in years)	3,01	4,935
Living in a small town (1 if in a small town, 0 otherwise)	−51,61	−3,778
Critical value of t-Statistic (at 5% significance level): 2,074		
R-squared: 0,634		
F statistic: 19,08		
Critical value of F statistic (at 5% significance level): 3,44		

Source: Authors

The interpretation of the obtained values of the parameters for the dummy variables should be conducted in relation to those variants of a given variable which are omitted in the final model. In our model, we initially considered three variants of place of living: living in a big city, living in a small town, and living in a rural area. However, only a dummy variable related to living in a small town turned out to be statistically significant. Hence, we should interpret the obtained parameter at this variable in relation to both alternative variants (i.e., living in a big city and living in a rural area). The negative value of the parameter for this dummy variable (Living in a small town) informs us that people living in the small towns tend to spend less (by USD 51,61 annually, on average) than people living in the big cities and people living in rural areas. Given that the related dummy variable (named Living in a big city) turned out to be statistically insignificant (and removed in the process of "general to specific modeling"), we can conclude that there is no statistically significant difference between people living in the big cities and people living in the rural areas in how much they spend (on average) on purchases of this drug.

As usual, economic knowledge and experience would be required for finding the reasons why people living in small towns tend to spend significantly less on purchases of the drug as compared to people living in big cities and people living in rural areas, as well as to find the reasons why there are no significant differences between amounts of spending on drug by people living in big cities and in villages.

5.2   Modeling Seasonality by Means of Dummy Variables

5.2.1   Introduction

Many economic processes are characterized by seasonality. Seasonality means that the level of variables presents some regularly recurring peaks, observable within the specified period. In an economic analysis the most common seasonal patterns are:

- Quarterly seasonality—when a variable shows peaks in some quarters and troughs in other quarters. For example: the sale of skiing equipment peaks in the first and fourth quarter of a year.
- Monthly seasonality—when a variable shows peaks in some months and troughs in other months. An example is the demand for holiday travel in Europe: it peaks in July and August.
- Daily seasonality—when a variable shows peaks in some days of a week and troughs in other days. Travel on regional trains is an example: regional trains are usually most crowded on Mondays and Fridays.

It is worth noting that the presence of clearly visible quarterly seasonality of a variable does not mean that there exists equally discernible monthly seasonality. For example, while sale of skiing equipment in continental Europe rises in the fourth quarter of a year and stays on high level in the first quarter of another year, it does not necessarily mean that analogous peaks of sale are observable in any specific months. This is so because snowfalls, although showing predictable quarterly seasonality, do not show equally predictable monthly seasonality. Sometimes winter is harsh and snow occurs as early as in October, so that the demand for skiing equipment jumps in this very October or early November, while in other years winter is much milder, with snowfalls occurring as late as just before Christmas. This means that the seasonal patterns may not be visible on a monthly basis, while they are clearly visible on a quarterly basis.

From the point of view of an econometric modeling, it is very important to distinguish between two different patterns of seasonality (Levenbach and Cleary 2006):

- Additive seasonality—when seasonal nominal differences of a variable from its trend show stable values.
- Multiplicative seasonality—when seasonal percentage deviations of a variable from its trend show stable values.

5.2.2   The Nature and Dealing with Additive Seasonality

Suppose that you are analyzing the quarterly cost of energy consumption in a company in the years 2007–2012. These data are presented in Table 5.4 as well as in Fig. 5.1. As can be seen, the investigated variable shows discernible quarterly seasonality. The consumption of energy tends to peak in the first and fourth quarters, while being on lower levels in the second and third quarters.

From the visual inspection of Fig. 5.1, we cannot infer accurately whether actual energy consumption deviates from its long-term trend in an additive or multiplicative way. The simplest way to check for this is to check whether the absolute values of nominal differences and absolute values of the percentage deviations of a variable from its trend are stable, not showing their own trends. Table 5.5 presents the actual values of the investigated variable, the fitted values obtained from its linear trend, absolute nominal differences between actual and fitted values, and absolute percentage deviations of actual values from fitted values. Figure 5.2 shows

Table 5.4 Quarterly cost of energy consumption (in thousands of USD) in a company in 2007–2012 years (with additive seasonality)

Year	1st quarter	2nd quarter	3rd quarter	4th quarter
2007	10.700	8.000	9.300	10.500
2008	14.100	10.800	10.000	12.900
2009	14.700	12.600	11.800	14.900
2010	17.500	14.200	15.400	15.700
2011	19.500	16.800	16.500	19.300
2012	21.500	18 .400	18.600	20.900
Average	16.333	13.467	13.600	15.700

Source: Authors

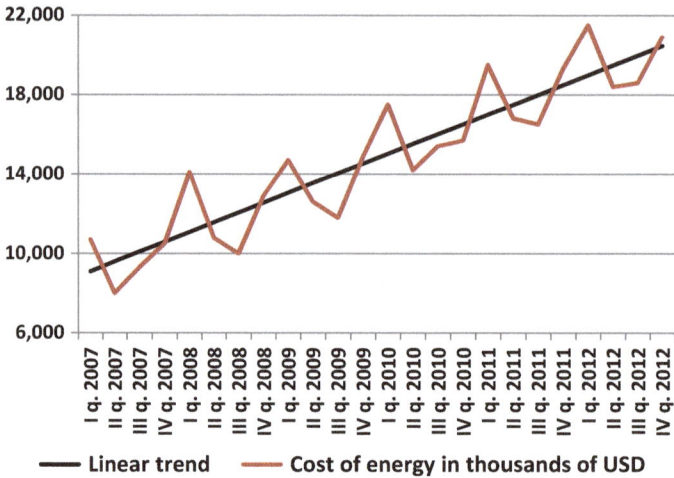

Fig. 5.1 Quarterly cost of energy consumption (in thousands of USD) in a company in 2007–2012 years (on the basis of data from Table 5.4). Source: Authors

the absolute nominal differences between the actual and fitted values of cost of energy consumption and a linear trend of these differences, while Fig. 5.3 presents the absolute percentage deviations of its actual values from its fitted values and a linear trend of these deviations.

The simple rules useful in differentiating between additive and multiplicative seasonality are following:

- If the value of R-squared of a linear trend of absolute percentage deviations of actual values of a variable from its trend is larger than the R-squared of a linear trend of absolute nominal differences of actual values from a trend, then the investigated variable is additively seasonal.
- If the value of R-squared of a linear trend of absolute nominal differences of actual values of a variable from its trend is larger than the R-squared of a linear trend of absolute percentage deviations of actual values from a trend, then the investigated variable is multiplicatively seasonal.

As we can infer from Figs. 5.2 and 5.3, the linear trend of absolute percentage deviations has an R-squared value of 0,718 and is much stronger than the linear trend of absolute nominal differences, which has R-square of 0,071. We conclude, therefore, that the quarterly cost of energy consumption has additive pattern of seasonality.

How can we deal with such additive seasonality in econometric modeling? One of many available techniques is to use seasonal dummy variables. Similarly as in the case of dummy variables for qualitative factors (discussed in Sect. 5.1), the main principle is to use $k - 1$ seasonal variables if you consider k seasons in the regression with intercept. Therefore, if we have quarterly data then we should consider three seasonal dummy variables. If we have monthly data, then we should

Table 5.5 Computation of absolute differences and absolute percentage deviations of actual cost of energy consumption from its linear trend (on the basis of data from Table 5.4)

Period	Quarterly cost of energy consumption (in thousands of USD)	Fitted values from linear trend	Absolute differences between actual and fitted values	Absolute percentage deviations of actual values from fitted values (%)
I q. 2007	10.700	9.104,00	1.596,00	17,5
II q. 2007	8.000	9.597,13	1.597,13	16,6
III q. 2007	9.300	10.090,26	790,26	7,8
IV q. 2007	10.500	10.583,39	83,39	0,8
I q. 2008	14.100	11.076,52	3.023,48	27,3
II q. 2008	10.800	11.569,65	769,65	6,7
III q. 2008	10.000	12.062,78	2.062,78	17,1
IV q. 2008	12.900	12.555,91	344,09	2,7
I q. 2009	14.700	13.049,04	1.650,96	12,7
II q. 2009	12.600	13.542,17	942,17	7,0
III q. 2009	11.800	14.035,30	2.235,30	15,9
IV q. 2009	14.900	14.528,43	371,57	2,6
I q. 2010	17.500	15.021,57	2.478,43	16,5
II q. 2010	14.200	15.514,70	1.314,70	8,5
III q. 2010	15.400	16.007,83	607,83	3,8
IV q. 2010	15.700	16.500,96	800,96	4,9
I q. 2011	19.500	16.994,09	2.505,91	14,7
II q. 2011	16.800	17.487,22	687,22	3,9
III q. 2011	16.500	17.980,35	1.480,35	8,2
IV q. 2011	19.300	18.473,48	826,52	4,5
I q. 2012	21.500	18.966,61	2.533,39	13,4
II q. 2012	18.400	19.459,74	1.059,74	5,4
III q. 2012	18.600	19.952,87	1.352,87	6,8
IV q. 2012	20.900	20.446,00	454,00	2,2

Source: Authors

consider eleven seasonal dummy variables. It does not mean, of course, that all these seasonal variables will be present in the final model, because some of them may turn out to be statistically insignificant.

If we have four seasons, we begin with considering three seasonal dummy variables. In modeling additive seasonality, each of these dummy variables has the value of unity in case of a given season and zero otherwise. So, for our data regarding quarterly cost of energy consumption, we consider the following seasonal dummy variables for additive seasonality:

- First Quarter—dummy variable having value of unity in the first quarter and zero in remaining three quarters.
- Second Quarter—dummy variable having value of unity in the second quarter and zero in remaining three quarters.

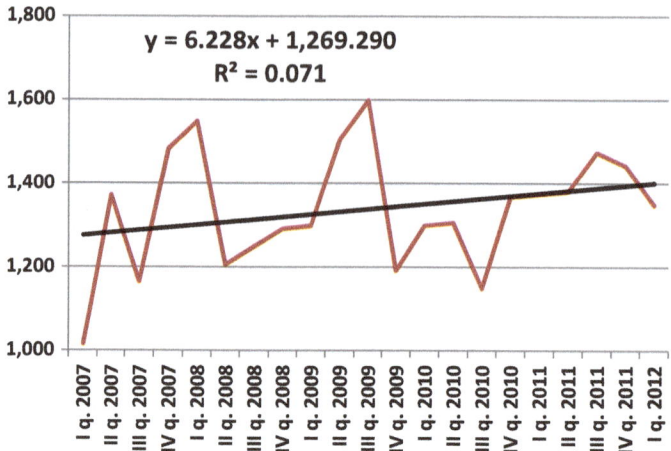

Fig. 5.2 The absolute nominal differences between actual and fitted values of cost of energy consumption (on the basis of data from Table 5.5). Source: Authors

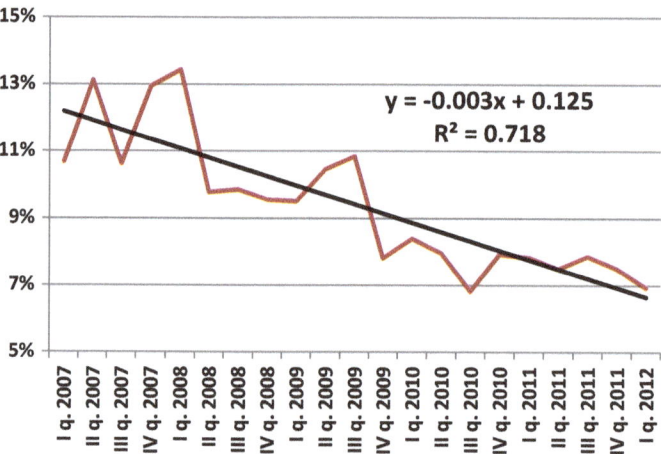

Fig. 5.3 The absolute percentage deviations of actual values of cost of energy consumption from its fitted values (on the basis of data from Table 5.5). Source: Authors

- Third Quarter—dummy variable having value of unity in the third quarter and zero in remaining three quarters.

The data for the estimation of trend regression with additive seasonality are presented in Table 5.6.

Now we estimate the parameters of the general model of quarterly cost of energy consumption against all four candidate explanatory variables, including the trend variable and the three seasonal dummies. The results of this regression are presented in Table 5.7.

Table 5.6 Data for the estimation of trend regression with additive seasonality

Period	Quarterly cost of energy consumption (in thousands of USD)	Trend variable	First quarter (dummy variable)	Second quarter (dummy variable)	Third quarter (dummy variable)
I q. 2007	10.700	1	1	0	0
II q. 2007	8.000	2	0	1	0
III q. 2007	9.300	3	0	0	1
IV q. 2007	10.500	4	0	0	0
I q. 2008	14.100	5	1	0	0
II q. 2008	10.800	6	0	1	0
III q. 2008	10.000	7	0	0	1
IV q. 2008	12.900	8	0	0	0
I q. 2009	14.700	9	1	0	0
II q. 2009	12.600	10	0	1	0
III q. 2009	11.800	11	0	0	1
IV q. 2009	14.900	12	0	0	0
I q. 2010	17.500	13	1	0	0
II q. 2010	14.200	14	0	1	0
III q. 2010	15.400	15	0	0	1
IV q. 2010	15.700	16	0	0	0
I q. 2011	19.500	17	1	0	0
II q. 2011	16.800	18	0	1	0
III q. 2011	16.500	19	0	0	1
IV q. 2011	19.300	20	0	0	0
I q. 2012	21.500	21	1	0	0
II q. 2012	18.400	22	0	1	0
III q. 2012	18.600	23	0	0	1
IV q. 2012	20.900	24	0	0	0

Source: Authors

Table 5.7 Results of a general regression of quarterly cost of energy consumption (in thousands of USD) against all four candidate explanatory variables (estimated on the basis of data from Table 5.6)

Variable	Coefficient	t-Statistic
Intercept	8.545,00	27,099
Trend variable	511,07	31,705
First quarter (dummy variable)	2.166,55	6,874
Second quarter (dummy variable)	−1.211,19	−3,868
Third quarter (dummy variable)	−1.588,93	−5,095

Critical value of t-Statistic (at 5% significance level): 2,093
R-squared: 0,984
F statistic: 284,17
Critical value of F statistic (at 5% significance level): 2,90

Source: Authors

As can be seen in Table 5.7, the estimated general regression of quarterly cost of energy consumption against all four candidate explanatory variables is statistically significant at the 5% level, and all structural parameters are also statistically significant at the same level. Therefore, we should abort the procedure of selecting explanatory variables and leave the model in its current form, so that the general model is the same as a specific model.

The interpretation of the obtained values of coefficients of the individual seasonal dummy variables should be conducted with respect to the season of the year which was omitted when creating these dummy variables. In our model, we used dummy variables for the first, second, and third quarter so we omitted the fourth one. This means that we should interpret the individual seasonal parameters as follows:

- In the first quarter of a year, the cost of energy consumption is, on average, higher than in the fourth quarter by USD 2.166,55 thousands.
- In the second quarter of a year, the cost of energy consumption is, on average, lower than in the fourth quarter by USD 1.211,19 thousands.
- In the third quarter of a year, the cost of energy consumption is, on average, lower than in the fourth quarter by USD 1.588,93 thousands.

If one or two out of these three seasonal dummies turned out to be statistically insignificant, then we should interpret the obtained parameters of the dummies which remained in the model in relation to those seasons which were omitted in the final model. For example, suppose that the dummy variable for the second quarter turned out to be statistically insignificant, leaving only First quarter and Third quarter in the final specific model, but the remaining parameters of the model are the same as presented in Table 5.7. In this case, we should conclude that there is no statistically significant difference resulting from seasonality between cost of energy consumption in the second and fourth quarter of a year. The remaining two parameters of the two seasonal dummies should be interpreted as follows:

- In the first quarter of a year, the cost of energy consumption is, on average, higher than in the second and fourth quarters by 2.166,55 thousands of USD.
- In the third quarter of a year, the cost of energy consumption is, on average, lower than in the second and fourth quarters by 1.588,93 thousands of USD.

5.2.3   The Nature and Dealing with Multiplicative Seasonality

The characteristic feature of the additive seasonality was the relative stability of nominal differences of actual values of a seasonal variable from their fitted values, as obtained from trend line or from a regression model without any seasonal variables. These differences from the trend line tended not to show any sharp trends. In contrast, the main feature of multiplicative seasonality is a relative stability of percentage deviations of actual values of a seasonal variable from their fitted values.

The additive seasonality in the above case of quarterly cost of energy consumption meant that in the first quarter of each year the energy consumption tended to be higher, on average, than in the fourth quarter, by 2.166,55 thousands of USD. So regardless of whether the actual energy consumption in the fourth quarter equaled 10.000 thousands of USD or 20.000 of USD, this deviation tended to stay constant in nominal terms. Although it stayed constant in nominal terms, it was not constant in terms of percentage deviations.

If the energy consumption in the fourth quarter was 10.000 thousands of USD, then in the following quarter it is expected to be 12.166,55 thousands of USD (21,7% higher than in the fourth quarter), and if the energy consumption in the fourth quarter was 20.000 thousands of USD, then in the following quarter it is expected to be 22.155,55 thousands of USD (10,8% higher than in the fourth quarter). The identical nominal difference between the fourth and the first quarter (i.e., 2.166,55 thousands of USD) entailed higher percentage deviation (21,7%) when the value for the fourth quarter was low (10.000 thousands of USD) and much lower percentage deviation (10,8%) when the value for the fourth quarter was higher (20.000 thousands of USD). That's why we could see in Fig. 5.3 the evidently declining trend of percentage deviations of actual values from fitted values. This was because the whole time series of our data was showing a rising trend throughout the whole sample.

In the case of multiplicative seasonality, the situation is just the reverse. The percentage deviations of actual observations from their fitted values tend to be stable while nominal deviations tend to show some evident trends. For example, it may be that in the first quarter of each year, the energy consumption tends to be higher, on average, than in the fourth quarter, by 20%. So regardless of whether the actual energy consumption in the fourth quarter equaled 10.000 thousands of USD or 20.000 of USD, the difference between the first and the fourth quarter tends to remain constant in percentage (not nominal) terms.

If the energy consumption in the fourth quarter was 10.000 thousands of USD, then in the following quarter it is expected to be 12.000 thousands of USD (20% higher than in the fourth quarter), and if the energy consumption in the fourth quarter was 20.000 thousands of USD, then in the following quarter it is expected to be 24.000 thousands of USD (again, 20% higher than in the fourth quarter). The identical percentage difference between the fourth and the first quarter (i.e., 20%) entailed smaller nominal difference (2.000 thousands of USD) when the value for the fourth quarter was small and much higher nominal difference (4.000 thousands of USD) when the value for the fourth quarter was higher. Although the difference tends to stay constant in percentage terms, it is not constant in terms of nominal differences. As a result, in the case of multiplicative seasonality, we can observe some evident trends of nominal differences of actual values from fitted values.

To illustrate this, suppose that now we have the data about the quarterly cost of energy consumption as shown in Table 5.8.

We can see that these quarterly data show evident seasonal patterns. The first quarters are featured by the highest energy consumption while the third quarters show the lowest energy consumption. However, from such raw data it is usually impossible, or very difficult, to draw inferences about the nature of this seasonality, whether it is additive or multiplicative. Sometimes we can get clear suggestions from the chart of these raw observations as shown in Fig. 5.4.

As can be seen in Fig. 5.4, the magnitude of nominal differences between actual cost of energy and its fitted values (from linear trend) increases with increasing values of the trend line. It immediately suggests the multiplicative nature of this seasonality. However, it is not always as clearly visible as in this case. Therefore, it is a good idea to check whether the absolute percentage deviations as well as absolute nominal differences between the actual values of a seasonal variable and

Table 5.8 Quarterly cost of energy consumption (in thousands of USD) in a company in 2007–2012 years (with multiplicative seasonality)

Year	1st quarter	2nd quarter	3rd quarter	4th quarter
2007	14.300	13.600	10.900	12.000
2008	23.900	20.000	11.100	18.800
2009	27.900	24.400	13.500	22.800
2010	36.300	28.400	19.500	24.400
2011	43.100	34.400	20.500	31.600
2012	49.900	38.400	23.500	34.800
Average	32.567	26.533	16.500	24.067

Source: Authors

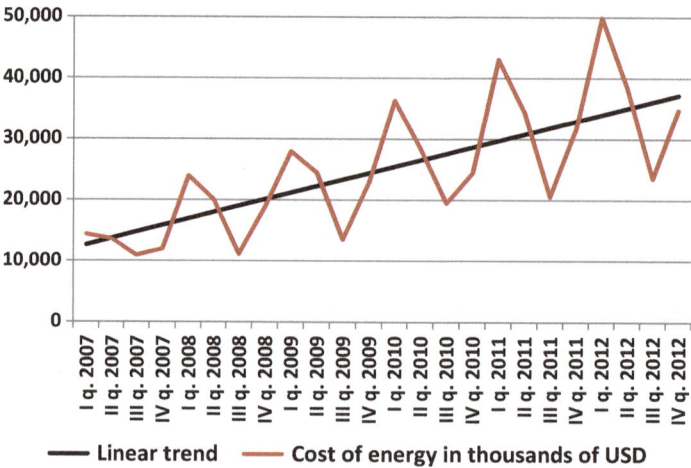

Fig. 5.4 Quarterly cost of energy consumption (in thousands of USD) in a company in 2007–2012 years (on the basis of data from Table 5.8). Source: Authors

Table 5.9 Computation of absolute differences and absolute percentage deviations of actual cost of energy consumption from its linear trend (on the basis of data from Table 5.8)

Period	Quarterly cost of energy consumption (in thousands of USD)	Fitted values from linear trend	Absolute differences between actual and fitted values	Absolute percentage deviations of actual values from fitted values (%)
I q. 2007	14.300	12.654,67	1.645,33	13,0
II q. 2007	13.600	13.720,93	120,93	0,9
III q. 2007	10.900	14.787,19	3.887,19	26,3
IV q. 2007	12.000	15.853,45	3.853,45	24,3
I q. 2008	23.900	16.919,71	6.980,29	41,3
II q. 2008	20.000	17.985,97	2.014,03	11,2
III q. 2008	11.100	19.052,23	7.952,23	41,7
IV q. 2008	18.800	20.118,49	1.318,49	6,6
I q. 2009	27.900	21.184,75	6.715,25	31,7
II q. 2009	24.400	22.251,01	2.148,99	9,7
III q. 2009	13.500	23.317,28	9.817,28	42,1
IV q. 2009	22.800	24.383,54	1.583,54	6,5
I q. 2010	36.300	25.449,80	10.850,20	42,6
II q. 2010	28.400	26.516,06	1.883,94	7,1
III q. 2010	19.500	27.582,32	8.082,32	29,3
IV q. 2010	24.400	28.648,58	4.248,58	14,8
I q. 2011	43.100	29.714,84	13.385,16	45,0
II q. 2011	34.400	30.781,10	3.618,90	11,8
III q. 2011	20.500	31.847,36	11.347,36	35,6
IV q. 2011	31.600	32.913,62	1.313,62	4,0
I q. 2012	49.900	33.979,88	15.920,12	46,9
II q. 2012	38.400	35.046,14	3.353,86	9,6
III q. 2012	23.500	36.112,41	12.612,41	34,9
IV q. 2012	34.800	37.178,67	2.378,67	6,4

Source: Authors

its fitted values present any clear trends. Table 5.9 contains the data prepared for such analysis.

Figure 5.5 shows the absolute nominal differences while Fig. 5.6 shows the absolute percentage deviations. As we can see, the trend of absolute nominal differences seems to be very strong and has a value of R-squared equaling 0,932. In contrast, the linear trend of absolute percentage deviations is much weaker (and has an R-squared of 0,047). We can conclude, therefore, that this particular time series of quarterly cost of energy consumption has a multiplicative pattern of seasonality.

How can we deal with such multiplicative seasonality in econometric modeling? Again, one of the available techniques is to use some seasonal dummy variables. As in other cases of dummy variables for qualitative factors, as well as in the case of dummies for the additive seasonality, the main principle is to use $k - 1$ seasonal

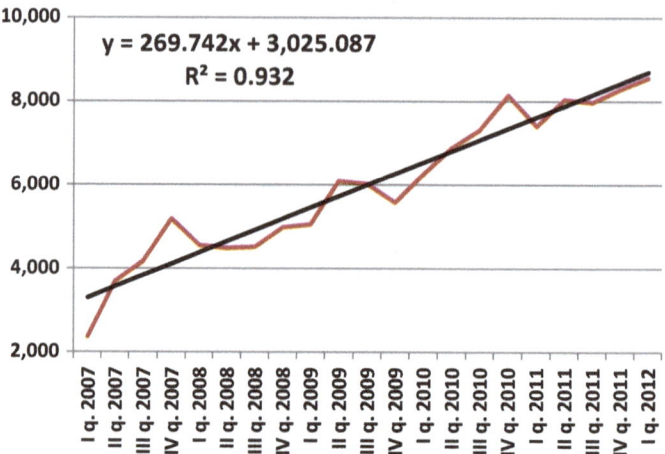

Fig. 5.5 The absolute nominal differences between actual and fitted values of cost of energy consumption (on the basis of data from Table 5.9). Source: Authors

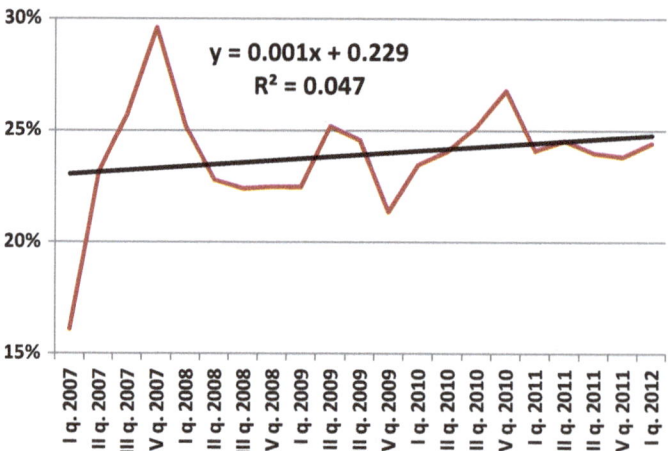

Fig. 5.6 The absolute percentage deviations of actual values of cost of energy consumption from its fitted values (on the basis of data from Table 5.9). Source: Authors

variables if you consider k seasons in the regression with intercept. Therefore, for our quarterly data, we should consider three seasonal dummy variables.

For our four seasons, we begin with three seasonal dummy variables. In modeling additive seasonality, each of these dummy variables had the value of unity in case of a given season and zero otherwise. However, in the case of multiplicative seasonality, the dummy variables look somewhat different. Instead of capturing seasonality in a binary way (unity vs. zero), in modeling multiplicative seasonality each of these dummy variables has a value equal to the number of the observation in case of a given season and zero otherwise. So, given that our time series of quarterly cost of energy consumption begins in the first quarter, we consider the following seasonal dummy variables for multiplicative seasonality:

- First Quarter—dummy variable having the following values: 1, 0, 0, 0, 5, 0, 0, 0, 9, . . .
- Second Quarter—dummy variable having the following values: 0, 2, 0, 0, 0, 6, 0, 0, 0, . . .
- Third Quarter—dummy variable having the following values: 0, 0, 3, 0, 0, 0, 7, 0, 0, . . .

The data for the estimation of trend regression with multiplicative seasonality are presented in Table 5.10.

Now we can estimate the parameters of the general model of quarterly cost of energy consumption against all four candidate explanatory variables: a trend

Table 5.10 Data for the estimation of trend regression with multiplicative seasonality

Period	Quarterly cost of energy consumption (in thousands of USD)	Trend variable	First quarter (dummy variable)	Second quarter (dummy variable)	Third quarter (dummy variable)
I q. 2007	14.300	1	1	0	0
II q. 2007	13.600	2	0	2	0
III q. 2007	10.900	3	0	0	3
IV q. 2007	12.000	4	0	0	0
I q. 2008	23.900	5	5	0	0
II q. 2008	20.000	6	0	6	0
III q. 2008	11.100	7	0	0	7
IV q. 2008	18.800	8	0	0	0
I q. 2009	27.900	9	9	0	0
II q. 2009	24.400	10	0	10	0
III q. 2009	13.500	11	0	0	11
IV q. 2009	22.800	12	0	0	0
I q. 2010	36.300	13	13	0	0
II q. 2010	28.400	14	0	14	0
III q. 2010	19.500	15	0	0	15
IV q. 2010	24.400	16	0	0	0
I q. 2011	43.100	17	17	0	0
II q. 2011	34.400	18	0	18	0
III q. 2011	20.500	19	0	0	19
IV q. 2011	31.600	20	0	0	0
I q. 2012	49.900	21	21	0	0
II q. 2012	38.400	22	0	22	0
III q. 2012	23.500	23	0	0	23
IV q. 2012	34.800	24	0	0	0

Source: Authors

Table 5.11 Results of a general regression of quarterly cost of energy consumption (in thousands of USD) against all four candidate explanatory variables (estimated on the basis of data from Table 5.10)

Variable	Coefficient	t-Statistic
Intercept	10.640,79	14,591
Trend variable	986,11	15,971
First quarter (dummy variable)	937,46	13,217
Second quarter (dummy variable)	313,42	4,593
Third quarter (dummy variable)	−482,69	−7,324

Critical value of t-Statistic (at 5% significance level): 2,093
R-squared: 0,978
F statistic: 210,77
Critical value of F statistic (at 5% significance level): 2,90

Source: Authors

variable as well as three seasonal dummies for multiplicative seasonality. The results of this regression are presented in Table 5.11.

As can be seen in Table 5.11, the estimated general regression of quarterly cost of energy consumption against all four candidate explanatory variables is statistically significant at the 5% level. All structural parameters are also statistically significant at the same level. Therefore, in this case we abort the procedure of selecting explanatory variables and leave the model in its current form.

The interpretation of the obtained values for the parameters that remain at the individual seasonal dummy variables should be again conducted in relation to this season of the year which was omitted when creating these dummy variables (the fourth quarter here). However, it is more complex than it was for the additive seasonality. We interpret the individual parameters obtained for seasonal dummies as follows:

- In the first quarter of a year, the cost of energy consumption is, on average, higher than in the fourth quarter by USD 937,46 thousands multiplied by the observation number in the sample (so $937,46 \times 1 = $ USD 937,46 thousands in the first quarter of 2007, then $937,46 \times 5 = $ USD 4.687,29 thousands in the first quarter of 2008, then $937,46 \times 9 = $ USD 8.437,12 thousands in the first quarter of 2009, and so on).

- In the second quarter of a year, the cost of energy consumption is, on average, lower than in the fourth quarter by USD 313,42 thousands multiplied by the observation number in the sample (so $313,42 \times 2 = $ USD 626,84 thousands in the second quarter of 2007, then $313,42 \times 6 = $ USD 1.880,53 thousands in the second quarter of 2008, then $313,42 \times 10 = $ USD 3.134,22 thousands in the second quarter of 2009, and so on).

- In the third quarter of a year, the cost of energy consumption is, on average, lower than in the fourth quarter by USD 482,69 thousands multiplied by the observation number in the sample (so $482,69 \times 3 = $ USD 1.448,08 thousands in the third quarter of 2007, then $482,69 \times 7 = $ USD 3.378,86 thousands in the third quarter of 2008, then $482,69 \times 11 = $ USD 5.309,63 thousands in the third quarter of 2009, and so on).

5.3 Using Dummy Variables for Outlying Observations

The application of dummy (binary) variables for the outlying observations was already illustrated in Chap. 3, where we identified two potential outliers in the dataset used in regression of quarterly cost of goods sold. The first outlier related to the 4th quarter of 2010 and the second one related to the 2nd quarter of 2011. Therefore, we extended our set of candidate explanatory variables by including the following two dummy variables:

- Dummy variable for the 4th quarter of 2010, denoted as $IVq2010$ and having the following values of observations:

$$IVq2010 = \begin{cases} 1 & \text{in the 4th quarter of 2010} \\ 0 & \text{otherwise} \end{cases}$$

- Dummy variable for the 2nd quarter of 2011, denoted as $IIq2011$ and having the following values of observations:

$$IIq2011 = \begin{cases} 1 & \text{in the 2nd quarter of 2011} \\ 0 & \text{otherwise} \end{cases}$$

In time-series regressions without lags as well as in cross-sectional regressions, the usage of dummy variables for outliers is as simple as in our regression of the quarterly cost of goods sold. We can deal with any outlier by adding an additional dummy variable with value of 1 in case of the outlying observation and 0 otherwise. However, the situation is more complex in the case of time-series models with lags. We will illustrate this with the example of simple linear autoregression of the fourth order.

Suppose that we intend to estimate the fourth-order autoregression of the growth of quarterly export sales of a product. Our modeled variable is denoted as ESG_t and represents the percentage growth of the export sales of a product in t-th quarter over the same quarter of the previous year. We have the quarterly data covering a 10-year period and resulting in 40 observations. Given that our intention is to estimate the fourth-order autoregression, our dataset looks as presented in Table 5.12. Figure 5.7 visualizes our dependent variable (that is, ESG_t).

Figure 5.7 shows that there are at least two significant outliers: the first quarter of 2006 (when the export sales fell by 87,6% y/y) and the first quarter of 2007 (when the growth of export sales equaled 710% y/y). These two observations are not independent from each other, because the more than sevenfold increase of export sales in the first quarter of 2007 results only from a so-called low base effect, stemming from very low export sales in the first quarter of 2006. If the volume of the export sales in the first quarter of 2005 amounted to, for example, 100 units, then

Table 5.12 Data for the fourth-quarter autoregression of a growth (y/y) of quarterly export sales of a product (denoted as ESG_t)

Period	ESG_t (%)	ESG_{t-1} (%)	ESG_{t-2} (%)	ESG_{t-3} (%)	ESG_{t-4} (%)
I q. 2003	21,5	–	–	–	–
II q. 2003	20,7	21,5	–	–	–
III q. 2003	18,3	20,7	21,5	–	–
IV q. 2003	15,1	18,3	20,7	21,5	–
I q. 2004	18,7	15,1	18,3	20,7	21,5
II q. 2004	13,8	18,7	15,1	18,3	20,7
III q. 2004	12,3	13,8	18,7	15,1	18,3
IV q. 2004	10,4	12,3	13,8	18,7	15,1
I q. 2005	5,3	10,4	12,3	13,8	18,7
II q. 2005	2,2	5,3	10,4	12,3	13,8
III q. 2005	−5,6	2,2	5,3	10,4	12,3
IV q. 2005	−8,3	−5,6	2,2	5,3	10,4
I q. 2006	−87,6	−8,3	−5,6	2,2	5,3
II q. 2006	−13,8	−87,6	−8,3	−5,6	2,2
III q. 2006	−17,5	−13,8	−87,6	−8,3	−5,6
IV q. 2006	−22,7	−17,5	−13,8	−87,6	−8,3
I q. 2007	710,0	−22,7	−17,5	−13,8	−87,6
II q. 2007	−25,2	710,0	−22,7	−17,5	−13,8
III q. 2007	−26,2	−25,2	710,0	−22,7	−17,5
IV q. 2007	−27,1	−26,2	−25,2	710,0	−22,7
I q. 2008	−25,4	−27,1	−26,2	−25,2	710,0
II q. 2008	−21,5	−25,4	−27,1	−26,2	−25,2
III q. 2008	−18,9	−21,5	−25,4	−27,1	−26,2
IV q. 2008	−17,2	−18,9	−21,5	−25,4	−27,1
I q. 2009	−12,2	−17,2	−18,9	−21,5	−25,4
II q. 2009	−10,8	−12,2	−17,2	−18,9	−21,5
III q. 2009	−10,0	−10,8	−12,2	−17,2	−18,9
IV q. 2009	−2,1	−10,0	−10,8	−12,2	−17,2
I q. 2010	−1,7	−2,1	−10,0	−10,8	−12,2
II q. 2010	1,5	−1,7	−2,1	−10,0	−10,8
III q. 2010	4,5	1,5	−1,7	−2,1	−10,0
IV q. 2010	7,5	4,5	1,5	−1,7	−2,1
I q. 2011	9,0	7,5	4,5	1,5	−1,7
II q. 2011	7,1	9,0	7,5	4,5	1,5
III q. 2011	7,2	7,1	9,0	7,5	4,5
IV q. 2011	6,9	7,2	7,1	9,0	7,5
I q. 2012	6,0	6,9	7,2	7,1	9,0
II q. 2012	6,6	6,0	6,9	7,2	7,1
III q. 2012	7,7	6,6	6,0	6,9	7,2
IV q. 2012	6,2	7,7	6,6	6,0	6,9

Source: Authors

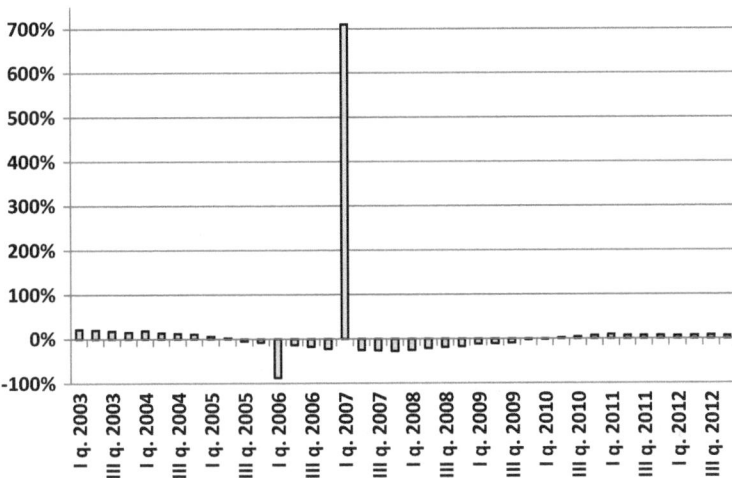

Fig. 5.7 The growth (y/y) of quarterly export sales of a product (denoted as ESG$_t$). Source: Authors

Table 5.13 Results of a fourth-order autoregression of a growth (y/y) of quarterly export sales of a product (denoted as ESG$_t$) without any adjustment for the outlying observations (estimated on the basis of data from Table 5.12)

Variable	Coefficient	t-Statistic
Intercept	0,20	0,906
ESG$_{t-1}$	−0,08	−0,476
ESG$_{t-2}$	−0,08	−0,452
ESG$_{t-3}$	−0,08	−0,441
ESG$_{t-4}$	−0,18	−0,997

Critical value of t-Statistic (at 5% significance level): 2,040
R-squared: 0,043
F statistic: 0,349
Critical value of F statistic (at 5% significance level): 2,68

Source: Authors

its volume in the first quarter of 2006 amounted to 12,4 units (after falling by 87,6% y/y). If that was the case, then the increase of export sales by 710% y/y in the first quarter of 2007 meant that the volume of sales in that quarter amounted to 100,4 units (after growing by 710% from 12,4 units). It means that this seemingly huge jump of sales in the first quarter of 2007 resulted only from the comeback of the sales to the level observed in the first quarter of 2005.

If we estimate the autoregression without any adjustment of the original data, then these two outlying quarters (first quarter of 2006 and first quarter of 2007) would have a disastrous impact on the regression results. As we can see in Table 5.13, the estimated autoregression is statistically insignificant at 5% significance level and has all parameters statistically insignificant.

The simplest and perhaps most tempting way of dealing with these two outlying quarters would be to remove them from the dataset. However, getting rid of the problems related to these outliers is not as simple as it might seem. This is so because of the lags of the explanatory variables. The data in Table 5.12 show that not only the first quarters of 2006 and 2007 are "infected" by these two outlying observations, but all the remaining quarters between the first quarter of 2006 and the first quarter of 2008! The only difference is that in the case of the first quarters of 2006 and 2007 the outliers are the observations of the dependent variable while in the case of the remaining seven "infected" quarters the outliers are the observations of the explanatory variables. Therefore, dropping these two outlying values would require dropping as many as nine consecutive observations, that is, all the observations in which these two outlying values occur as either the dependent variable or any of the explanatory variables. It means that in order to fully remove these two extreme values from the dataset we would have to get rid of the data from more than a 2-year period.

Our example is the autoregression of the fourth order. If we estimated the autoregression of the eighth order then the number of lost observations would increase significantly (even if there are only two outlying values of the dependent variable). Therefore, removing the observations "infected" by the outlying values is not a reasonable idea. In such cases, we should deal with the identified outliers in a different way.

The alternative way of treating outliers is to use dummy variables. We have considered two dummy variables in our regression of cost of goods sold, because only two observations seemed to be "infected" by the outlying values. However, the general principle is to include in the set of candidate explanatory variables as many dummy variables as the number of observations "infected" by the outlying values, although not all of them will turn out to be statistically significant in the estimated model. As was stated above, nine observations are "infected" by the outlying values of the modeled variable in our autoregression. These are the nine quarters between the first quarter of 2006 and the first quarter of 2008. Therefore, we should consider nine dummy variables for the outliers in our dataset. These dummy variables are presented in Table 5.14.

Finally, our set of the candidate explanatory variables for our autoregression, which now includes the dummy variables for the outliers, consists of thirteen variables: four lags of the dependent variable and nine dummy variables for the outliers. The results of the general regression are presented in Table 5.15.

As can be seen in Table 5.15, the general regression of a growth of quarterly export sales against all 13 candidate explanatory variables is statistically significant but has some structural parameters which seem to be statistically insignificant. Therefore, we reduce this general model by dropping those explanatory variables which are not statistically significant. According to "general to specific modeling" procedure, we excluded these variables in the following order: Dummy for IV q. 2006, Dummy for III q. 2006, ESG_{t-2}, Dummy for III q. 2007, ESG_{t-3}, Dummy for IV q. 2007. As a result, we obtained the specific model with seven explanatory variables. The results of its estimation are shown in Table 5.16.

As can be seen in Table 5.16, the obtained specific model has a very high value of its empirical F statistic (equaling 11.770,25) and includes the coefficients which are

Table 5.14 Data for the fourth-quarter autoregression of a growth (y/y) of quarterly export sales of a product (denoted as ESG_t)

Period	Dummy variable for:								
	I q. 2006	II q. 2006	III q. 2006	IV q. 2006	I q. 2007	II q. 2007	III q. 2007	IV q. 2007	I q. 2008
I q. 2004	0	0	0	0	0	0	0	0	0
II q. 2004	0	0	0	0	0	0	0	0	0
III q. 2004	0	0	0	0	0	0	0	0	0
IV q. 2004	0	0	0	0	0	0	0	0	0
I q. 2005	0	0	0	0	0	0	0	0	0
II q. 2005	0	0	0	0	0	0	0	0	0
III q. 2005	0	0	0	0	0	0	0	0	0
IV q. 2005	0	0	0	0	0	0	0	0	0
I q. 2006	1	0	0	0	0	0	0	0	0
II q. 2006	0	1	0	0	0	0	0	0	0
III q. 2006	0	0	1	0	0	0	0	0	0
IV q. 2006	0	0	0	1	0	0	0	0	0
I q. 2007	0	0	0	0	1	0	0	0	0
II q. 2007	0	0	0	0	0	1	0	0	0
III q. 2007	0	0	0	0	0	0	1	0	0
IV q. 2007	0	0	0	0	0	0	0	1	0
I q. 2008	0	0	0	0	0	0	0	0	1
II q. 2008	0	0	0	0	0	0	0	0	0
III q. 2008	0	0	0	0	0	0	0	0	0
IV q. 2008	0	0	0	0	0	0	0	0	0
I q. 2009	0	0	0	0	0	0	0	0	0
II q. 2009	0	0	0	0	0	0	0	0	0
III q. 2009	0	0	0	0	0	0	0	0	0
IV q. 2009	0	0	0	0	0	0	0	0	0

(continued)

Table 5.14 (continued)

Period	Dummy variable for:								
	I q. 2006	II q. 2006	III q. 2006	IV q. 2006	I q. 2007	II q. 2007	III q. 2007	IV q. 2007	I q. 2008
I q. 2010	0	0	0	0	0	0	0	0	0
II q. 2010	0	0	0	0	0	0	0	0	0
III q. 2010	0	0	0	0	0	0	0	0	0
IV q. 2010	0	0	0	0	0	0	0	0	0
I q. 2011	0	0	0	0	0	0	0	0	0
II q. 2011	0	0	0	0	0	0	0	0	0
III q. 2011	0	0	0	0	0	0	0	0	0
IV q. 2011	0	0	0	0	0	0	0	0	0
I q. 2012	0	0	0	0	0	0	0	0	0
II q. 2012	0	0	0	0	0	0	0	0	0
III q. 2012	0	0	0	0	0	0	0	0	0
IV q. 2012	0	0	0	0	0	0	0	0	0

Source: Authors

Table 5.15 Results of a general regression of a growth (y/y) of quarterly export sales of a product (denoted as ESG_t) against its four lags as well as nine dummy variables for the outlying observations (estimated on the basis of data from Tables 5.12 and 5.14)

Variable	Coefficient	t-Statistic
Intercept	0,00	−0,123
ESG_{t-1}	0,84	4,524
ESG_{t-2}	0,38	1,401
ESG_{t-3}	0,17	0,628
ESG_{t-4}	−0,50	−2,644
Dummy for I q. 2006	−0,76	−25,839
Dummy for II q. 2006	0,65	4,463
Dummy for III q. 2006	0,26	1,239
Dummy for IV q. 2006	0,08	0,398
Dummy for I q. 2007	6,95	47,667
Dummy for II q. 2007	−6,20	−4,534
Dummy for III q. 2007	−2,78	−1,405
Dummy for IV q. 2007	−1,30	−0,640
Dummy for I q. 2008	3,64	2,639

Critical value of t-Statistic (at 5% significance level): 2,074
R-squared: 1,000
F statistic: 6.592,77
Critical value of F statistic (at 5% significance level): 2,20

Source: Authors

Table 5.16 Results of a specific regression of a growth (y/y) of quarterly export sales of a product (denoted as ESG_t)

Variable	Coefficient	t-Statistic
Intercept	0,00	−0,682
ESG_{t-1}	1,25	17,215
ESG_{t-4}	−0,30	−4,869
Dummy for I q. 2006	−0,75	−28,059
Dummy for II q. 2006	0,96	13,919
Dummy for I q. 2007	7,12	151,608
Dummy for II q. 2007	−9,14	−17,497
Dummy for I q. 2008	2,20	4,865

Critical value of t-Statistic (at 5% significance level): 2,048
R-squared: 1,000
F statistic: 11.770,25
Critical value of F statistic (at 5% significance level): 2,36

Source: Authors

all statistically significant, except for the intercept. However, we remember from Table 5.13 that the fourth-order autoregression without any dummies for the outlying observations was statistically insignificant. Also, the R-squared of the specific regression equals unity, while in the fourth-order autoregression it equaled only 0,043. This is a good illustration of the extent to which neglecting the outliers can distort the regression results.

5.4   Dealing with Structural Changes in Modeled Relationships

5.4.1   Illustration of Structural Changes of Regression Parameters

Unfortunately for statisticians and economic analysts, the great majority of economic processes that include causal relationships between different economic variables are not stable in time. Instead, they are constantly changing. This relates to both macroeconomic processes as well as to sectoral, regional, and company-level relationships. This changing nature of most economic variables poses problems for statisticians (and particularly for forecasters), because it often makes modeling economic processes very troublesome (and sometimes impossible). Estimation of regression models assumes stability of parameters of the modeled relationships. If this condition is not satisfied, then regression analysis may result in obtaining false regression parameters, incorrect analytical conclusions about the structure of modeled relationships, and inaccurate forecasts, even if the model seems to be good within the sample (Evans 2003).

For changing relationships between variables included in the regression model, we will use the terms "structural changes" or "structural breaks." Because these changes occur either gradually or as one-off events with the passage of time, they must be always considered in time-series models. Ignoring significant structural changes may be very dangerous and may lead to false analytical findings. We will discuss the pitfalls of neglecting structural changes in Chap. 6.

We will illustrate one of the methods useful in dealing with structural breaks with the hypothetical example of the relationship between production expenses and production volume in a company. Suppose that you are a management accountant and your task is to track the causal relationship between changes of production volume as explanatory variable and changes of total production expenses as dependent variable.

Suppose that we have data on monthly total production costs (in thousands of USD) and monthly production volume (in physical units, e.g., kilograms) from 24 consecutive months. Your task is to separate, on the basis of such data, the fixed and variable elements in the cost structure. Fixed costs are those which do not change with shifting production level. In contrast, variable costs tend to follow changes in production volume. Knowing the structure of operating costs (i.e., what is the approximate percentage of fixed costs in total costs) is very important for managing a company, particularly in cyclical industries (i.e., industries with unstable sales volume). We will discuss this in more detail in the next chapter.

Suppose that we have the data presented in Table 5.17. However, from the accounting department, we can receive only the data presented in the last two columns: monthly production volume in physical units and total monthly production costs. The department of financial accounting in our company does not separate total operating costs into their fixed and variable parts. This separation is the duty of

Table 5.17 Monthly total production costs (in thousands of USD) and monthly production volume (in physical units) in 24 consecutive months in one company

Number of period	Monthly fixed expenses[a]	Unit variable cost[b]	Monthly production volume in physical units	Total monthly production costs[c]
1	1.000	2,00	1.000	3.000
2	1.050	1,97	1.500	4.005
3	1.100	1,94	2.000	4.980
4	1.150	1,91	1.800	4.588
5	1.200	1,88	1.200	3.456
6	1.250	1,85	1.500	4.025
7	1.300	1,82	1.400	3.848
8	1.350	1,79	1.000	3.140
9	1.400	1,76	800	2.808
10	1.450	1,73	1.100	3.353
11	1.500	1,70	1.400	3.880
12	1.550	1,67	1.200	3.554
13	1.650	1,64	1.100	3.454
14	1.750	1,61	1.600	4.326
15	1.850	1,58	2.200	5.326
16	1.950	1,55	2.000	5.050
17	2.050	1,52	1.700	4.634
18	2.150	1,49	1.000	3.640
19	2.250	1,46	1.300	4.148
20	2.350	1,43	1.100	3.923
21	2.450	1,40	800	3.570
22	2.550	1,37	1.200	4.194
23	2.650	1,34	1.400	4.526
24	2.750	1,31	1.000	4.060
Arithmetic mean	1.737,50	1,66	1.345,83	3.978,67
Median	1.600,00	1,66	1.250,00	3.964,00

Source: authors
[a]In thousands of USD per month
[b]In thousands of USD per month per one physical unit of production
[c]Monthly fixed expenses + (unit variable cost × monthly production volume in physical units)

the department of management accounting. Therefore, we have at our disposal only the aggregate data from the last two columns. However, the preceding two columns, monthly fixed expenses and unit variable costs, present the true but unobservable data regarding the structure of production costs.

For example, in the first month the fixed costs amounted to 1.000 thousands of USD while the variable cost per unit of production was USD 2 thousand. This means that the increase (decrease) of production level by one unit resulted in an increase (decrease) of total production costs by USD 2 thousands, on average. At a monthly production volume of 1.000 units, the total production costs in that period equaled USD 3.000 thousands $(1.000 + (2,00 \times 1.000) = 3.000)$. In contrast, in the last month, the fixed costs amounted to USD 2.750 thousands while the variable cost per unit of production amounted to USD 1,31 thousands, which means that increase (decrease) of production level by one unit in that period resulted in increase (decrease) of total production costs by USD 1,31 thousands, on average. At monthly production volume equaling 1.000 units, the total production costs in the last period equaled USD 4.060 thousands $(2.750 + (1,31 \times 1.000) = 4.060)$.

As we can infer from the data presented in columns "Monthly fixed expenses" and "Unit variable cost" (which are unobservable and are underlying the observable data presented in the last column), the structure of production costs of this company was gradually but significantly changing in the investigated 24-month period. Particularly, the nominal fixed costs increased by USD 50 thousand with each single month, while the nominal variable costs per unit of production decreased by USD 0,03 thousands every month. As a result, the structure of costs was getting "more fixed." These are conclusions which can be easily drawn from the unobservable data on "Monthly fixed expenses" and "Unit variable cost." They cannot be so easily inferred from the observable data from the last two columns of Table 5.17.

In many books on management accounting and cost accounting, we can find the discussion of simple regression as an objective tool of estimating fixed and variable costs on the basis of observable data regarding total costs and total production. Such regression usually has the following form:

$$TC_t = \alpha_0 + \alpha_1 TP_t,$$

where:

TC_t—total production costs in period t,

TP_t—total production volume (physical units or other cost drivers) in period t,

α_0—intercept, interpreted as average periodic fixed costs (i.e., costs which do not change with changing production volume),

α_1—slope parameter, interpreted as average variable production costs per unit of production (i.e., increase/decrease of total production costs caused by increase/decrease of production volume by one unit).

We now regress the total monthly production costs, in the last column of Table 5.17, against monthly production volume. This relationship is presented in Fig. 5.8 and in Table 5.18.

As can be seen in Table 5.18, the regression of monthly total production costs against monthly production volume is statistically significant and has two structural parameters which are both statistically significant. However, the R-squared of this regression, although seemingly high (0,798), is understated. This is so because our dependent variable (monthly total production costs) is generated in a fully deterministic way as the linear function of our explanatory variable (monthly production volume).

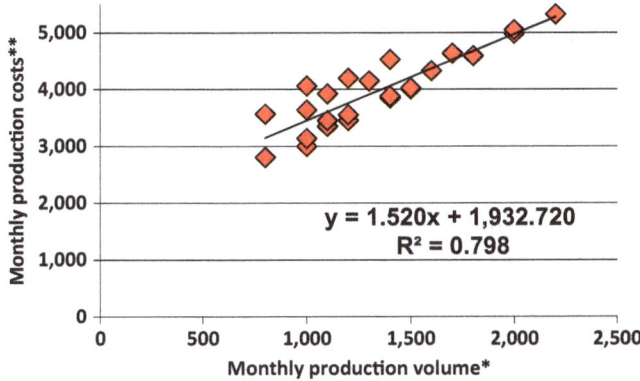

Fig. 5.8 Relationship between monthly total production costs (in thousands of USD) and monthly production volume (in physical units) in 24-month period (on the basis of data from the last two columns of Table 5.17). *In physical units of production. **In thousands of USD per month. Source: Authors

Table 5.18 Results of linear regression of monthly total production costs (in thousands of USD and denoted as TC_t) against monthly production volume (denoted as TP_t), estimated on the basis of data from the last two columns of Table 5.17

Variable	Coefficient	t-Statistic
Intercept	1.932,72	8,474
TP_t	1,52	9,310
Critical value of t-Statistic (at 5% significance level): 2,074		
R-squared: 0,798		
F statistic: 86,68		
Critical value of F statistic (at 5% significance level): 4,30		

Source: Authors

If the relationship between two variables is linear and deterministic, then R-squared should equal unity. The reason why deterministic data-generating process applied here resulted in R-squared significantly lower than unity is the gradually changing structure of parameters (i.e., gradually rising monthly fixed expenses and gradually falling unit variable cost). However, not only R-squared in distorted (which is not a big problem) but also both structural parameters have misleading values, which is much more problematic. We will address this issue in more detail in Chap. 6.

Structural changes like this can be partially dealt with in several different ways. Here, we will discuss only one of the available procedures, based on dummy variables for structural changes. Interested readers may find more comprehensive discussions of methods dealing with structural breaks in other books.

5.4.2 Dealing with Structural Changes by Means of Dummy Variables

To test for the presence of structural breaks in the modeled relationship, and to capture them in the estimated regression, we apply the following procedure, called sequential test for structural change (Table 5.19).

We check the presence of structural changes, by means of the procedure described above, in the regression of monthly total production costs against monthly production volume (shown in Table 5.18). The tested regression was estimated on the basis of monthly time-series data from 24-month period and in

Table 5.19 The procedure for testing and modeling the structural changes with the use of dummy variables

Step 1	Formulate the hypotheses: H_0:parameters of the estimated regression are stable (i.e., there are not any significant structural changes in modeled relationships) H_1:parameters of the estimated regression are not stable (i.e., there are some significant structural changes in modeled relationships)
Step 2	Estimate $t-1$ auxiliary regressions, each with all the variables from the tested regression and one additional dummy variable of the following form: $D_i = (0_0, \ldots, 0_{i-1}, 1_i, \ldots, 1_t)$, where: t—number of observations in tested regression, i—potential location (within time series of data) of the structural change of the relationship modeled by tested regression (changing from $i=1$ to $i=t-1$)
Step 3	For all $t-1$ auxiliary regressions estimated above, compute the empirical statistic F_i according to the following formula: $F_i = \frac{R_{2i}^2 - R_1^2}{1 - R_{2i}^2}(n - (k+1))$, where: R_{2i}^2—coefficient of determination (R-squared) of i-th auxiliary regression estimated in Step 2 (i.e., tested regression with added auxiliary variable D_i), R_1^2—coefficient of determination (R-squared) of the tested regression, n—number of observations on which the tested model was estimated, k—number of explanatory variables in the tested model
Step 4	From $t-1$ auxiliary regressions estimated in Step 2 select the one with the highest value of F_i (and denote it as F_{imax})
Step 5	Obtain the critical (theoretical) value F_i^* from the Fisher–Snedecor table, for $m_1 = 2$, $m_2 = n - (k+1)$ and for assumed significance level α (usually assumed at 5%)
Step 6	Compare F_{imax} with critical F_i^* If $F_{imax} \leq F_i^*$, then H_0 is not rejected, which means that there are no significant structural changes in the relationships modeled by tested regression If $F_{imax} > F_i^*$., then H_0 is rejected, which means that there are some significant structural changes in the relationships modeled by tested regression
Step 7	If H_0 is rejected in Step 6, then leave the respective auxiliary dummy variable D_i (i.e., D_i identified as having F_{imax}) in the tested regression (that is, include D_i in the set of its explanatory variables of your model)

Source: Authors

Number of period	D_1	D_2	D_3	...	D_{21}	D_{22}	D_{23}
1	1	1	1	...	1	1	1
2	0	1	1	...	1	1	1
3	0	0	1	...	1	1	1
4	0	0	0	...	1	1	1
5	0	0	0	...	1	1	1
6	0	0	0	...	1	1	1
7	0	0	0	...	1	1	1
8	0	0	0	...	1	1	1
9	0	0	0	...	1	1	1
10	0	0	0	...	1	1	1
11	0	0	0	...	1	1	1
12	0	0	0	...	1	1	1
13	0	0	0	...	1	1	1
14	0	0	0	...	1	1	1
15	0	0	0	...	1	1	1
16	0	0	0	...	1	1	1
17	0	0	0	...	1	1	1
18	0	0	0	...	1	1	1
19	0	0	0	...	1	1	1
20	0	0	0	...	1	1	1
21	0	0	0	...	1	1	1
22	0	0	0	...	0	1	1
23	0	0	0	...	0	0	1
24	0	0	0	...	0	0	0

Table 5.20 Auxiliary dummy explanatory variables (D_i) created for the regression of monthly total production costs against monthly production volume (estimated on 24 observations and presented in Table 5.18)

Source: Authors

our case $t = 24$. This means that we need to create $t - 1 = 23$ auxiliary explanatory variables and we need to estimate $t - 1 = 23$ auxiliary regressions. The way of creating these $t - 1 = 23$ auxiliary explanatory variables is illustrated in Table 5.20.

We now estimate $t - 1 = 23$ auxiliary regressions and compute the F_i statistic for each of these regressions. Our tested model was estimated on the basis of 24 observations (so $n = 24$) and included one explanatory variable (so $k = 1$). The obtained values of F_i statistics for individual auxiliary variables D_i are presented in Table 5.21. The example of a computation of F_i for D_1 is as follows:

$$F_i = \frac{R_{2i}^2 - R_1^2}{1 - R_{2i}^2}\left(n - (k + 1)\right) = \frac{0,820 - 0,798}{1 - 0,820}\left(24 - (1 + 1)\right) = 2,80.$$

According the information provided in the last row of Table 5.21, the maximum value of F_i statistic (F_{imax}) equals 146,00. This was obtained for D_{17}. Now we compare the obtained F_{imax} with critical value F_i^*. For $n = 24$ and $k = 1$, we obtain:

Table 5.21 Results of auxiliary regressions estimated for potential structural changes in the relationship between monthly total production costs and monthly production volume

Number of auxiliary explanatory variable D_i	R_{2i}^2	R_1^{2a}	F_i
D_1	0,820	0,798	2,80
D_2	0,822	0,798	3,05
D_3	0,815	0,798	2,04
D_4	0,816	0,798	2,15
D_5	0,827	0,798	3,70
D_6	0,834	0,798	4,80
D_7	0,843	0,798	6,36
D_8	0,858	0,798	9,34
D_9	0,877	0,798	14,34
D_{10}	0,894	0,798	20,16
D_{11}	0,907	0,798	26,13
D_{12}	0,925	0,798	37,60
D_{13}	0,942	0,798	54,97
D_{14}	0,948	0,798	63,41
D_{15}	0,951	0,798	68,60
D_{16}	0,962	0,798	96,44
D_{17}	0,973	0,798	146,00
D_{18}	0,956	0,798	78,79
D_{19}	0,946	0,798	59,81
D_{20}	0,923	0,798	36,02
D_{21}	0,889	0,798	18,00
D_{22}	0,863	0,798	10,48
D_{23}	0,839	0,798	5,59
Maximum value of F_i (F_{imax}) =			146,00

Source: Authors
[a]R-squared from the regression of monthly total production costs against monthly production volume (presented in Table 5.18)

$$m_1 = 2,$$

$$m_2 = n - (k + 1) = 24 - (1 + 1) = 22.$$

For the level of statistical significance set at $\alpha = 0,05$, and determined values of m_1 and m_2, we obtain the critical value of F_i^* from the Fisher–Snedecor table, which equals 3,44.

Finally, we compare F_{imax} with the value of critical F_i^*. In this case $F_{imax} > F_i^*$, which means that H_0 is rejected, and we conclude that there were some significant structural changes in the relationships between monthly total production costs and monthly production volume. Therefore, the test for the structural breaks confirms what we already knew: gradually changing fixed and variable costs. Given that H_0

Table 5.22 Results of linear regression of monthly total production costs against monthly production volume (denoted as TC_t) and dummy variable for structural change (denoted as D_{17})

Variable	Coefficient	t-Statistic
Intercept	1.970,52	23,309
TP_t	1,83	27,758
D_{17}	−640,35	−11,805

Critical value of t-Statistic (at 5% significance level): 2,080
R-squared: 0,973
F statistic: 146,00
Critical value of F statistic (at 5% significance level): 3,47

Source: Authors

was rejected, we leave the respective auxiliary dummy variable (i.e., D_{17}) in the set of explanatory variables in the regression of monthly total production costs against monthly production volume.

After running this procedure for identification of structural changes, the complete model including a dummy variable as a proxy for structural change is presented in Table 5.22. As can be seen, the whole regression is strongly significant and includes three statistically significant structural parameters. We also note that the goodness of fit statistic (R-squared) now equals 0,973 and is much higher than in the original regression (where it equaled 0,798).

5.4.3 Dangers of Interpreting Individual Structural Parameters from the Models with Dummy Variables for Structural Changes

Some words of caution are important here. First, although inclusion of dummy variable for structural change resulted in significant improvement of goodness of fit, manifested in boosted R-squared, we still must be careful when interpreting structural parameters of the model. From the bottom part of Table 5.17, we know that true monthly fixed costs in the investigated period had the arithmetic average of USD 1.737,50 thousand and median value of USD 1.600 thousand. But these fixed costs were gradually rising from month to month and at the end of the sample they amounted to USD 2.750 thousand per month. Meanwhile, the intercept in both our regressions is similar and equals 1.932,72 in the regression without D_{17} and 1.970,52 in the regression with D_{17}. Therefore, both intercepts heavily underestimate true fixed costs from the most recent months, which are usually most important for economic analysis and company management, and significantly overstate the median and average monthly fixed costs within the whole sample. The intercept from the extended regression (i.e., regression with D_{17}) is close only to the actual monthly fixed costs from the period for which the structural break was detected, but that period is rather meaningless for current economic analysis.

Better approximation of current (i.e., at the end of the time-series data) monthly fixed production costs is obtained by combining the intercept and parameter for D_{17}. The structural parameter obtained for D_{17} was negative and equaled $-640,35$. Given that the observations of D_{17} equaled unity from the beginning of the sample up to the seventeenth observation and zero further onward, we conclude that the actual production costs in the subsample from eighteenth observation to the end of the sample exceeded actual costs in the preceding subsample by 640,35 thousand of USD, on average. If we assume that this difference can be allocated to the fixed costs, then we can obtain the estimate of the average fixed costs in the subsample from eighteenth observation onward equaling 2.610,87 (1.970,52 + 640,35), which is much closer to the actual monthly fixed costs at the end of the sample (which equaled 2.750 thousand of USD).

However, if we look at the slope parameter from the two regressions (which are often interpreted as variable production costs per unit of production), we can see that its value equaled 1,52 in the reduced model (i.e., model without D_{17}) and 1,83 in the extended model (i.e., model with D_{17}). Meanwhile, from the bottom part of Table 5.17, we know that both median and arithmetic average of true unit production costs equaled 1,66. So, while the intercept from the reduced model understated the average actual variable costs (by 8,4%), the intercept from the extended model overstated these actual costs by 10,2%. However, the unit variable production costs were gradually declining throughout the whole analyzed period, from USD 2,00 per unit in the first month to USD 1,31 per unit in the last month of the sample. If the analyst or manager is interested in having an estimate of the current (recent) variable costs per unit, for uses such as pricing decisions or for computations of break-even point, then the reduced model overstates the recent true variable costs by about 16% (1,52/1,31) and the extended model overstates the true costs by almost 40% (1,83/1,31)! So, the extended model, despite having much better goodness of fit to actual data than the reduced model, produces much more heavily biased estimate of current unit variable costs. Making economic decisions on the basis of such heavily distorted numbers may be very hazardous and disastrous!

Another important caveat relates to the interpretation of the structural break. Our dummy variable was constructed in a binary way, assuming that there is only one structural break in the whole sample. In our case, this structural break was identified to occur between seventeenth and eighteenth observation. However, the actual structural changes in our example were gradual and lasted from the beginning to the end of the sample, instead of being centered around any single observation. Therefore, it is usually unjustified to treat the observation at which the dummy variable changes value (from unity to zero) as the location of the structural break. If the actual structural changes are spread in time, then using dummy variables to locate the time of structural break is a significant abuse. Only if we have the knowledge that the structural break had a one-off but permanent nature (e.g., as a result of replacing the old production line by a new one), then we can interpret the values of observations of a dummy variable for structural break.

5.4.4 Testing for Structural Changes in Our Model

We now check the presence of any structural breaks in our model for cost of goods sold. Our model was estimated on the basis of 20 quarterly observations, so in our case $t = 20$. This means that we need to create $t - 1 = 19$ auxiliary explanatory variables and we need to estimate $t - 1 = 19$ auxiliary regressions. Then, for each of these 19 regressions, we need to compute the F_i statistic. Our model of quarterly cost of goods sold was estimated on the basis of 20 observations ($n = 20$) and included three explanatory variables (so $k = 3$). The obtained values of F_i statistics for individual auxiliary variables D_i are presented in Table 5.23.

According the information provided in the last row of Table 5.23, the maximum value of F_i statistic (F_{imax}) equals 8,82. This was obtained for D_{17}. Now we need to compare the obtained F_{imax} with the critical value F_i^*. For $n = 20$ and $k = 3$, we obtain:

$$m_1 = 2,$$

$$m_2 = n - (k + 2) = 20 - (3 + 1) = 16.$$

Table 5.23 Results of auxiliary regressions estimated for our model of costs of goods sold

Number of auxiliary explanatory variable D_i	R_{2i}^2	R_1^{2a}	F_i
D_1	0,994	0,993	1,81
D_2	0,993	0,993	1,02
D_3	0,993	0,993	0,47
D_4	0,993	0,993	0,19
D_5	0,993	0,993	0,19
D_6	0,993	0,993	1,53
D_7	0,993	0,993	0,86
D_8	0,994	0,993	1,78
D_9	0,994	0,993	4,07
D_{10}	0,993	0,993	0,26
D_{11}	0,993	0,993	0,04
D_{12}	0,993	0,993	1,12
D_{13}	0,994	0,993	2,79
D_{14}	0,993	0,993	0,00
D_{15}	0,993	0,993	0,08
D_{16}	0,994	0,993	1,84
D_{17}	0,995	0,993	8,82
D_{18}	0,995	0,993	7,32
D_{19}	0,994	0,993	3,72
Maximum value of F_i (F_{imax}) =			8,82

Source: Authors

For the level of statistical significance set at $\alpha = 0,05$, and determined values of m_1 and m_2, we obtain the critical value of F_i^* from the Fisher–Snedecor table, which equals 3,55.

Finally, we compare F_{imax} with critical F_i^*. In the case of our model $F_{imax} > F_i^*$, which means that H_0 is rejected, and we conclude that there are some significant structural changes in the relationships quantified by our model. Perhaps, the problem with autocorrelation of residuals detected in Chap. 4 was caused by such structural changes. Therefore, it is suggested to adjust our model, by adding to the set of its explanatory variables the dummy variable capturing the structural break (D_{17}). After such adjustment of the model, the full process of model estimation and verification should be conducted once again.

5.5   Dealing with In-Sample Non-linearities

5.5.1   Relevance of a Functional Form of a Regression Model

One of the steps in verifying the correctness of the estimated linear model is checking for the correctness of its general specification by means of Ramsey's *RESET* test. One of the causes of a misspecification of a linear regression is its incorrect functional form (e.g., linear instead of some nonlinear function). The incorrect functional form of a regression model may result in obtaining residuals which are strongly autocorrelated or heteroscedastic. Moreover, such misspecified models often tend to generate biased predictions of a dependent variable. Therefore, if the Ramsey's *RESET* test suggests that the linear model is misspecified, it is justified to consider some alternative functional forms.

In this section, we will illustrate the estimation procedures for two commonly applied nonlinear forms of regression functions: a power regression and an exponential regression. Readers interested in nonlinear models can find many other alternative nonlinear functional forms discussed in more advanced books on econometrics.

5.5.2   Estimating Power Regression

Suppose that we are building the regression explaining a company's monthly production output in units of products (denoted as P) as the function of two explanatory variables: monthly input of machine hours (denoted as M) and monthly input of labor hours (denoted as L). This will be called a production function, and it will assume the power functional form. Such a nonlinear (power) output model is known as the Cobb–Douglas production function. Table 5.24 presents the actual data of the model variables (P, M and L) from the last 10 months.

Table 5.24 Actual data of the dependent variable (P) and two explanatory variables (M and L)

Number of month	P (monthly production output in units)	M (monthly input of machine hours)	L (monthly input of labor hours)
1	75.000	1.000	300
2	70.000	800	350
3	84.000	1.200	250
4	80.000	1.100	300
5	60.000	900	200
6	55.000	700	250
7	57.000	1.000	200
8	80.000	1.000	300
9	110.000	1.400	400
10	70.000	1.100	250

Source: Authors

If we consider the power functional form of this regression, the theoretical model looks as follows:

$$P = \alpha_0 \, M^{\alpha_1} \, L^{\alpha_2},$$

where α_0, α_1, α_2 are the structural parameters.

The ordinary least squares (OLS) method of estimating parameters of the model assumes linearity of the relationship between dependent variable and the model parameters. Therefore, if we assume the power functional form of our production function, we cannot apply the ordinary least squares directly to the original data presented in Table 5.24 above. Instead, we have to do some artificial transformation of the data, which we will call a linearization.

Mathematically, the procedure of linearization runs as follows. If we take the logarithm of both sides of the above power function, we obtain:

$$\log P = \log(\alpha_0 \, M^{\alpha_1} \, L^{\alpha_2}).$$

The logarithm of a product is the equivalent of the addition of the logarithms. Therefore, we transform the right-hand side of the above formula as follows:

$$\log P = \log \alpha_0 + \log M^{\alpha_1} + \log L^{\alpha_2}.$$

Further, the logarithms of the powers of M and L can be transformed as follows:

$$\log P = \log \alpha_0 + \alpha_1 \log M + \alpha_2 \log L.$$

We now introduce the following notation:

$$\log P = \text{LP},$$
$$\log M = \text{LM},$$
$$\log L = \text{LL},$$
$$\log \alpha_0 = a.$$

After such a series of transformations, we have obtained the function of the following form:

$$LP = a + \alpha_1 \, LM + \alpha_2 \, LL.$$

As we see, this is a linear function of LP (which is the auxiliary variable derived from logging the original dependent variable P) against two explanatory variables: LM (the auxiliary variable derived from logging the original explanatory variable M) and LL (the auxiliary variable derived from logging the original explanatory variable L). In this transformed (linearized) regression, there are also two structural parameters from the original power function (α_1, α_2) as well as one new structural parameter a (the auxiliary parameter derived from logging the original intercept α_0). This transformed regression can now be estimated by the ordinary least squares. However, before we estimate this regression, we have to transform the original observations of P, M, and L by computing their logarithms (in order to obtain the values of LP, LM, and LL). The results of such logarithmic transformations of the dependent and explanatory variables are shown in Table 5.25.

The results of regressing LP against LM and LL are presented in Table 5.26. Our transformed (linearized) regression has the following form:

$$LP = 1,570 + 0,666 \, LM + 0,531 \, LL.$$

Now we can return from this transformed regression to the original one with a power functional form. As we remember, in the transformed (linearized) regression, there are two structural parameters from the original power function (α_1, α_2). Their values can be transferred directly to the power function. However, in the transformed regression, there is also an auxiliary structural parameter a, obtained by logging the original intercept α_0. Therefore, before we transfer this parameter to the original power function, we have to "de-log" it.

Table 5.25 Transformed (logged) data of the dependent variable (P) and two explanatory variables (M and L), calculated on the basis of the data from Table 5.24

Number of month	LP = log P	LM = log M	LL = log L
1	4,8751	3,0000	2,4771
2	4,8451	2,9031	2,5441
3	4,9243	3,0792	2,3979
4	4,9031	3,0414	2,4771
5	4,7782	2,9542	2,3010
6	4,7404	2,8451	2,3979
7	4,7559	3,0000	2,3010
8	4,9031	3,0000	2,4771
9	5,0414	3,1461	2,6021
10	4,8451	3,0414	2,3979

Source: Authors

Table 5.26 The results of regressing LP against LM and LL (on the basis of the data from Table 5.25)

Variable	Coefficient	t-Statistic
Intercept	1,570	5,459
LM $= \log M$	0,666	7,214
LL $= \log L$	0,531	6,502
R-squared: 0,951		

Source: Authors

If:

$$\log\alpha_0 = a = 1,570,$$

then

$$\alpha_0 = 10^a = 10^{1,570} = 37,142.$$

Finally, our power production function looks as follows:

$$P = 37,142 \ M^{0,666} \ L^{0,531}.$$

In the above power function, the slope parameters at the two explanatory variables have the following interpretation:

- The increase/decrease of the input of machine hours by 1% causes the increase/decrease of the total production output by 0,666% on average (when keeping the input of the labor hours unchanged).
- The increase/decrease of the input of labor hours by 1% causes the increase/decrease of the total production output by 0,531% on average (when keeping the input of the machine hours unchanged).

5.5.3 Estimating Exponential Regression

We illustrate the process of estimating exponential regressions on the basis of the same data which were used in the above discussion of the power function. Suppose again that we are building the regression explaining a company's monthly production output in units of products (denoted as P) as the function of two explanatory variables: monthly input of machine hours (denoted as M) and monthly input of labor hours (denoted as L).

If we now consider the exponential functional form of this regression, the theoretical model looks as follows:

$$P = \alpha_0 \ \alpha_1{}^M \ \alpha_2{}^L,$$

where α_0, α_1, α_2 are the structural parameters.

Again, we cannot apply the ordinary least squares directly to the original data (presented in Table 5.24). Instead, we have do some artificial transformation of the data (a linearization). If we log both sides of the above exponential function, we obtain:

$$\log P = \log\left(\alpha_0 \, \alpha_1{}^M \, \alpha_2{}^L\right).$$

Applying the properties of the logarithm, we transform the right-hand side of the above formula as follows:

$$\log P = \log \alpha_0 + \log \, \alpha_1{}^M + \log \, \alpha_2{}^L.$$

Further, the logarithms of the powers of α_1 and α_2 can be transformed as follows:

$$\log P = \log \alpha_0 + M \log \alpha_1 + L \log \alpha_2.$$

We now introduce the following notation:

$$\begin{aligned} \log P &= \text{LP}, \\ \log \alpha_0 &= a, \\ \log \alpha_1 &= b, \\ \log \alpha_2 &= c. \end{aligned}$$

After such a series of transformations, we have obtained the function of the following form:

$$\text{LP} = a + b \, M + c \, L.$$

As we see, LP (the auxiliary variable derived from logging the original dependent variable P) is now a linear function of the parameters and of the two original explanatory variables: M and L. In this transformed (linearized) regression, there are three structural parameters: a, b, and c (the auxiliary parameters derived from logging the original parameters α_0, α_1, and α_2). This transformed regression can now be estimated by the ordinary least squares. However, before we estimate this regression, we have to transform the original observations of P by computing their logarithms (in order to obtain the values of LP). However, in contrast to the procedure of estimating a power function, here we do not have to do any transformation of the original values of explanatory variables (M and L). The transformed (logged) observations of the dependent variable as well as the original observations of the explanatory variables are shown in Table 5.27.

The results of regressing LP against M and L are presented in Table 5.28.

So our transformed (linearized) regression has the following form:

$$\text{LP} = 4,35813 + 0,00029 \, M + 0,00075 \, L.$$

Now we return from this transformed regression to the original one (which has an exponential functional form). As we remember, in the transformed (linearized) regression, there are three auxiliary structural parameters (a, b, and c) obtained by logging the original structural parameters (α_0, α_1, and α_2, respectively). Therefore,

Table 5.27 Transformed data for the exponential regression of P against M and L (on the basis of the data from Table 5.24)

Number of month	LP = log P	M	L
1	4,8751	1.000	300
2	4,8451	800	350
3	4,9243	1.200	250
4	4,9031	1.100	300
5	4,7782	900	200
6	4,7404	700	250
7	4,7559	1.000	200
8	4,9031	1.000	300
9	5,0414	1.400	400
10	4,8451	1.100	250

Source: Authors

Table 5.28 The results of regressing LP against M and L (on the basis of the data from Table 5.27)

Variable	Coefficient	t-Statistic
Intercept	4,35813	86,764
M	0,00029	6,104
L	0,00075	5,114

R-squared: 0,937

Source: Authors

before we transfer these parameters to the original exponential function, we have to "de-log" them as follows:

- If $\log\alpha_0 = a = 4,35813$ then $\alpha_0 = 10^a = 10^{4,35813} = 22.810,1651$,
- If $\log\alpha_1 = b = 0,00029$ then $\alpha_1 = 10^b = 10^{0,00029} = 1,0007$,
- If $\log\alpha_2 = c = 0,00075$ then $\alpha_2 = 10^c = 10^{0,00075} = 1,0017$.

Finally, our exponential production function looks as follows:

$$P = 22.810,1651\ M^{1,0007}\ L^{1,0017}.$$

In the above power function, the slope parameters of the two explanatory variables have the following interpretation:

- The increase/decrease of the input of machine hours by 1 h causes the increase/ decrease of the total production output by 0,07% on average (when keeping the input of the labor hours unchanged).
- The increase/decrease of the input of labor hours by 1 h causes the increase/ decrease of the total production output by 0,17% on average (when keeping the input of the machine hours unchanged).

References

Evans MK (2003) Practical business forecasting. Blackwell, Oxford

Johnston J (1984) Econometric methods. McGraw-Hill, New York

Levenbach H, Cleary JP (2006) Forecasting. Practice and process for demand management. Thomson-Brooks/Cole, Belmont

Pindyck RS, Rubinfeld DL (1998) Econometric models and economic forecasts. Irwin McGraw-Hill, Boston

Chapter 6
Common Pitfalls in Regression Analysis

6.1 Introduction

Much too often the analytical tools offered by statistics and econometrics can be heavily abused. The incorrect application of the regression analysis and unskilled interpretation of its results might result in making disastrous economic decisions. In these circumstances, quantitative analysis can produce highly misleading results. In the following sections of this chapter, we will illustrate some common pitfalls of the reckless regression analysis.

6.2 Distorting Impact of Multicollinearity on Regression Parameters

In Chap. 4, we illustrated the distorting effect which is exerted on t-Statistics by the multicollinearity of explanatory variables. As serious a problem as it can be, in our opinion it is not the most dangerous pitfall related to the multicollinearity of explanatory variables. This is so because it is usually accompanied by a clear symptom of some problem, in the form of the contradictory combination of statistically significant regression (large F statistic as well as relatively large R-squared) with statistically insignificant individual explanatory variables (small absolute values of their t-Statistics).

There is another relevant pitfall related to multicollinearity. This seems to be more serious because it may occur also when the relationship between individual explanatory variables is not very strong though statistically significant. This is typically present (but not limited to) in autoregressive models (i.e., models in which lags of dependent variable are included in the set of explanatory variables). We illustrate this with the numerical example.

© Springer International Publishing AG 2018
J. Welc, P.J.R. Esquerdo, *Applied Regression Analysis for Business*,
https://doi.org/10.1007/978-3-319-71156-0_6

Imagine that we intend to build a simple regression by which we could evaluate the expected business conditions in one of the industries with a predictive horizon of up to four quarters. Suppose that we have time-series data about the pace of growth (year-over-year) of the industrial production in that sector. Clearly, the faster this pace of growth, the better are the business conditions in the industry. The simplest way of foreseeing short-term future is to predict the expected business data on the grounds of their current and past values. But to make such predictions, we need to estimate the model in which the forecasted variable is the dependent variable and its own lags constitute the set of explanatory variables. Such models are called autoregressions. If we consider regressing the annual growth of sectoral industrial production in t-th quarter Y_t against its past, lagged values, we may have the set of data presented in Table 6.1.

The more the current business conditions (proxied by sectoral industrial production) are correlated with their past (lagged) observations, the larger is the extent to which we can evaluate probable business conditions in the nearest future on the basis of these past observations. So we check the extent to which current growth of sectoral industrial production (Y_t) is correlated with its own observations lagged up to the fourth quarter. Table 6.2 presents the results of regressing Y_t against its lags, separately for each lag (simple regressions).

As can be seen, all four regressions are statistically significant at 5% significance level and have statistically significant explanatory variables. Although slope parameters, R-squared, and F statistics gradually and monotonically decline when lagging Y_t further into the past (from Y_1 to Y_4), we can clearly see that there exists significant "inertia" in the analyzed dependent variable (i.e., growth of sectoral industrial production). Even the relationship between Y_t and Y_{t-4} seems to be quite strong in light of the interval of time of 1 year, between the dependent variable (Y_t) and the explanatory variable (Y_{t-4}).

Turning to the interpretation of slope parameters, we can see that they have positive values in all four regressions. Although their values gradually decline with extension of the lags of dependent variable (which informs us that the "inertia" effect steadily fades away), we can clearly see that for this data set, all four regressions are informative about future business conditions in the industry.

Particularly, the statistical significance of the regression of Y_t against Y_{t-4}, combined with the positive value of its slope parameter, seems to be very useful. It gives us the possibility of forecasting the business conditions in our industry as far into the future as 1 year from now, on the grounds of current business conditions. Given that the intercept of this regression is statistically insignificant, we can treat it as being equal to zero. Therefore, we could make simplified forecasts of sectoral industrial production growth with a four quarter horizon just on the basis of the slope coefficient. For example, if the industrial production is now rising by 10% year-over-year, we can expect that four quarters from now this will slow down to about 6.7% year-over-year (10% times the slope coefficient of 0.67). Generally, from the values of slope parameters in all four regressions, we can conclude that the faster the industrial production is growing now, the faster (although slower than now) it is expected to grow in the course of the next four quarters.

Table 6.1 The quarterly data related to annual growth[a] of sectoral industrial production and its lags

Number of quarter	Y_t (%)	Y_{t-1} (%)	Y_{t-2} (%)	Y_{t-3} (%)	Y_{t-4} (%)
1	21,5	–	–	–	–
2	20,7	21,5	–	–	–
3	18,3	20,7	21,5	–	–
4	15,1	18,3	20,7	21,5	–
5	18,7	15,1	18,3	20,7	21,5
6	13,8	18,7	15,1	18,3	20,7
7	12,3	13,8	18,7	15,1	18,3
8	10,4	12,3	13,8	18,7	15,1
9	5,3	10,4	12,3	13,8	18,7
10	2,2	5,3	10,4	12,3	13,8
11	−5,6	2,2	5,3	10,4	12,3
12	−8,3	−5,6	2,2	5,3	10,4
13	−8,3	−8,3	−5,6	2,2	5,3
14	−13,8	−8,3	−8,3	−5,6	2,2
15	−17,5	−13,8	−8,3	−8,3	−5,6
16	−22,7	−17,5	−13,8	−8,3	−8,3
17	−25,1	−22,7	−17,5	−13,8	−8,3
18	−25,2	−25,1	−22,7	−17,5	−13,8
19	−26,2	−25,2	−25,1	−22,7	−17,5
20	−27,1	−26,2	−25,2	−25,1	−22,7
21	−25,4	−27,1	−26,2	−25,2	−25,1
22	−21,5	−25,4	−27,1	−26,2	−25,2
23	−18,9	−21,5	−25,4	−27,1	−26,2
24	−17,2	−18,9	−21,5	−25,4	−27,1
25	−14,5	−17,2	−18,9	−21,5	−25,4
26	−10,8	−14,5	−17,2	−18,9	−21,5
27	−10,0	−10,8	−14,5	−17,2	−18,9
28	−2,1	−10,0	−10,8	−14,5	−17,2
29	−1,7	−2,1	−10,0	−10,8	−14,5
30	1,5	−1,7	−2,1	−10,0	−10,8
31	4,5	1,5	−1,7	−2,1	−10,0
32	7,5	4,5	1,5	−1,7	−2,1
33	9,0	7,5	4,5	1,5	−1,7
34	7,1	9,0	7,5	4,5	1,5
35	7,2	7,1	9,0	7,5	4,5
36	6,9	7,2	7,1	9,0	7,5
37	6,0	6,9	7,2	7,1	9,0
38	6,6	6,0	6,9	7,2	7,1
39	7,7	6,6	6,0	6,9	7,2
40	6,2	7,7	6,6	6,0	6,9

Source: Authors
[a]((Industrial production in a given quarter/Industrial production in the corresponding quarter of the preceding year) − 1) × 100%

Table 6.2 Results of simple regressions of sectoral industrial production (Y_t) against its individual lags of up to fourth quarter

Specification of a regression[a]	Intercept[b]	Slope parameter[b]	R-squared	F statistic
Y_t against Y_{t-1}	0,00 (−0,802)	0,95 (24,264)	0,945	588,74
Y_t against Y_{t-2}	−0,01 (−1,108)	0,88 (13,865)	0,850	192,24
Y_t against Y_{t-3}	−0,02 (−1,374)	0,78 (9,233)	0,715	85,25
Y_t against Y_{t-4}	−0,02 (−1,566)	0,67 (6,566)	0,559	43,11

Critical value of t-Statistic (at 5% significance level): 2,032
Critical value of F statistic (at 5% significance level): 4,13

Source: Authors
[a]All four regressions estimated on the basis of data from periods from 5 through 40 (36 quarters), that is excluding the first four quarters lost when making lags of dependent variable
[b]Empirical t-Statistics in brackets

But if all four regressions are statistically significant individually, then why not combine all four lags into a single multivariate autoregressive model which would presumably better capture the relationship between current and lagged industrial production than any of the above simple univariate autoregressions. We begin with the general model in which current industrial production is the dependent variable while all its four lags constitute the set of candidate explanatory variables. Then we apply the procedure of "general to specific modeling" to remove the redundant lags. This general model is shown in Table 6.3.

Here, the estimated general model is strongly statistically significant, given its F statistic. However, not all the explanatory variables seem to be statistically significant. So it is recommended to reduce our general model by dropping the redundant explanatory variables. In accordance with the procedure of "general to specific modeling," we first remove Y_{t-3}, because of its lowest absolute value of t-Statistic. In the next step, after reestimating the regression without Y_{t-3}, it turns out that also Y_{t-2} is statistically insignificant. Finally, there are two explanatory variables remaining in the specific model, presented in Table 6.4.

Now both explanatory variables are statistically significant at the 5% significance level. So we can proceed to the interpretation of individual structural parameters. But there is a puzzle here, because the coefficient at Y_{t-4} is negative and statistically significant! From Table 6.2, we know that in the simple regression of Y_t against Y_{t-4}, the slope parameter was positive, statistically significant, and equaled 0,67! So, what is the direction of the relationship between current growth of sectoral production and its growth four quarters ahead? Positive or negative? Well, it seems that the negative value of structural parameter staying at Y_{t-4} is counterintuitive and probably distorted.

However, not only the coefficient of Y_{t-4} seems strange. When we look at the slope parameter at Y_{t-1}, we can see that it has the value of 1,20 while in the simple regression of Y_t against Y_{t-1} the slope parameter equaled only 0,95. The difference

Table 6.3 Results of general model of sectoral production growth year-over-year (Y_t) against its own four lags (of up to fourth quarter), estimated on the basis of data from Table 6.1

Variable	Coefficient	t-Statistic
Intercept	0,00	−0,550
Y_{t-1}	0,99	5,996
Y_{t-2}	0,30	1,243
Y_{t-3}	0,04	0,148
Y_{t-4}	−0,40	−2,477

Critical value of t-Statistic (at 5% significance level): 2,040
R-squared: 0,972
F statistic: 266,99
Critical value of F statistic (at 5% significance level): 2,68

Source: Authors

Table 6.4 The specific model of sectoral production growth year-over-year (Y_t) against its statistically significant lags

Variable	Coefficient	t-Statistic
Intercept	0,00	−0,542
Y_{t-1}	1,20	21,249
Y_{t-4}	−0,27	−5,195

Critical value of t-Statistic (at 5% significance level): 2,035
R-squared: 0,970
F statistic: 532,91
Critical value of F statistic (at 5% significance level): 3,28

Source: Authors

is rather large and complicates the economic interpretation. In the autoregressive model, the value of parameter below unity suggests that the modeled autoregressive process is self-correcting. The parameter equaling 0,95 suggests that if the production is currently growing relatively fast by, say, 10% year-over-year, in the following quarter it will be about 9,5% year-over-year (10% × 0,95), in the next quarter it will be about 9% (9,5% × 0,95), and so on. So, it will gradually revert to more natural long-term pace of growth.

Likewise, if the industry faces deep recession and the production is now declining by, say, 15% year-over-year, then the parameter of 0,95 suggests that it will be gradually reverting to the "normal" levels with every consecutive quarter and the declines in the following quarters will be shallower (i.e., −14,2% year-over-year in the next quarter, −13,5% year-over-year in the another quarter, and so on). Meanwhile, the value of slope parameter equaling 1,20 means that the autoregressive process is self-destabilizing, instead of being self-correcting! It suggests that current quarter growth by 10% year-over-year is followed by growth of 12% year-over-year (10% × 1,20) in the next quarter, 14,4% year-over-year (12% × 1,20) in another quarter, and so on (forever!). Clearly this is not possible and suggests that something is wrong here!

The problem is that when explanatory variables are correlated, then all the structural parameters can be heavily distorted *individually*, in possible different directions. Here, the slope parameter at Y_{t-1} is overstated, with a value of 1,20, while the slope parameter at Y_{t-4} is understated, with a value of −0,27). In such a case, the interpretation of individual parameters makes no economic sense, even though they are statistically significant. Instead, we should interpret the structural parameters jointly, by looking at the sum of their individual values.

Here, the sum of both structural parameters equals 0,93 (1,20–0,27) and suggests the following interpretation:

- The direction of the relationship between current growth of sectoral production and its lagged growth is generally positive, regardless of negative values of some individual slope parameters. This means that the faster is the pace of growth now, the faster is the expected growth in the forthcoming few quarters.
- The process is self-correcting, because the sum of all autoregressive parameters is smaller than unity (0,93). This suggests that if the pace of production growth now is relatively high/low as compared to long-term average, we should expect that in the following quarters the forecast for this growth will gradually decrease/increase toward more "normal" average levels.

Despite the individual slope parameters being biased in different directions, the sum of these parameters is much more sensible and as a result the whole model may be quite good in predicting future sectoral production. The overstatement of one slope parameter is to a large extent "smoothed-out" by the understatement of another parameter. Therefore, the distortion of individual parameters does not automatically mean that the whole model is useless. The only thing is that we cannot sensibly interpret the parameters at individual explanatory variables.

The problem of individual slope parameters being distorted by multicollinearity does not occur only in case of time-series models. It may be also present in many cross-sectional models. For example, when regressing relative market values of stock market companies against their financial results, we would expect that market capitalizations are positively related to companies' profitability (the higher profitability, the higher the market value of stock). But which profitability? In the income statement of every company we have different tiers of profits, including gross profit on sales, operating profit, gross profit before extraordinary items, net earnings, net earnings attributable to controlling shareholders, among others. According to the financial statement analysis books, each of these profit numbers carries some important and incremental information about company's ability to generate profits. However, profitability ratios based on profits reported on different levels of income statement tend to be positively correlated with each other. As a result, it may turn out that in the regression of companies' values against the set of profitability ratios, some of the individual slope parameters are negative, while the sum of all slope parameters related to profitability is positive.

Finally, returning to our autoregressive model of sectoral industrial production growth, we should mention the reasons why we knew that both explanatory variables (i.e., Y_{t-1} as well as Y_{t-4}) were correlated. If we turn back to Table 6.2, we can see that the simple regression of Y_t against Y_{t-3} was statistically significant (having R-squared of 0,715 and F statistic of 85,25). Because the time interval between dependent variable and lagged dependent variable in that simple regression equals three quarters, this is the same as the time interval

Table 6.5 Results of variance inflation factor test run for explanatory variables from the specific model

Auxiliary regression	R-squared	VIF_i statistic
Y_{t-1} against Y_{t-4}	0,728	2,67
Critical value of VIF_i statistic (as a "rule of thumb"): 5,00		

Source: Authors

between both explanatory variables in our specific model (i.e., between Y_{t-1} and Y_{t-4}). Therefore, the regression of Y_t against Y_{t-3}, presented in Table 6.2, is the same as the regression of Y_{t-1} against Y_{t-4}. Thus, if the relationship between Y_t and Y_{t-3} is statistically significant, then the relationship between Y_{t-1} and Y_{t-4} must be statistically significant as well, which means that there is some multicollinearity of these explanatory variables.

Unfortunately, this obvious multicollinearity of both explanatory variables may be overlooked by common tools applied in testing for multicollinearity. In Chap. 4, we presented the popular test which is variance inflation factor test. The results of running this test are presented in Table 6.5.

As we can see, the linear regression of one explanatory variable (Y_{t-1}) against another explanatory variable (Y_{t-4}) has coefficient of determination of 0,728 and variance inflation factor equaling 2,67. This VIF_i statistic lies clearly below the commonly applied subjective threshold of 5, which suggests that there is not any significant multicollinearity between Y_{t-1} and Y_{t-4}. Hence, in this case, the application of variance inflation factor test would result in overlooking the problem of both explanatory variables being correlated. This is a good illustration of how important it is not to apply any statistical tools (including tests for model correctness) mechanically, because such mechanical application can often bring dangerous consequences and misinterpretation of the obtained results. It is always crucial in statistical analysis to use common sense combined with economic theory and practical experience.

6.3 Analyzing Incomplete Regressions

One of the most common abuses of econometric modeling is estimating and interpreting the individual parameters of incomplete regressions. These are the regressions in which the set of explanatory variables does not include all the relevant explanatory variables. In such cases, the estimated values of the individual slope parameters often deviate heavily from their true values.

We illustrate this problem with a very simple numerical example. Imagine a dataset of values for the dependent variable Y and its eight deterministic drivers (causes), denoted by $X_1, \ldots, X_8$. These data are presented in Table 6.6. In this data-generating process, the observations on the dependent variable Y constitute the simple arithmetic sums of the values of all eight explanatory variables (causes).

Table 6.6 The dataset of the dependent variable (Y) and its eight deterministic causes (denoted as $X_1, \ldots, X_8$)

Number of observation	Y^*	X_1	X_2	X_3	X_4	X_5	X_6	X_7	X_8
1	26	1	1	1	1	5	8	1	8
2	39	7	6	4	6	4	4	2	6
3	42	2	8	5	4	7	6	3	7
4	38	2	3	5	5	6	5	7	5
5	39	4	5	4	7	5	7	3	4
6	38	2	1	9	4	8	3	1	10
7	32	3	1	7	5	4	6	1	5
8	37	1	5	4	9	5	5	1	7
9	35	5	3	1	6	4	8	2	6
10	30	1	8	1	3	3	5	4	5
11	50	5	7	8	3	7	8	4	8
12	45	4	6	7	4	6	7	5	6
13	51	6	7	8	3	7	9	4	7
14	49	5	6	7	4	8	8	6	5
15	45	7	5	6	5	7	6	5	4
16	54	6	6	7	6	6	7	6	10
17	48	5	5	6	5	7	8	7	5
18	46	6	6	5	4	5	7	6	7
19	40	5	4	6	5	3	6	5	6
20	38	4	3	7	4	4	5	6	5

Source: Authors

[a]$Y = X_1 + X_2 + X_3 + X_4 + X_5 + X_6 + X_7 + X_8$

Therefore, the true relationship between the dependent variable and the full set of explanatory variables is as follows:

$$Y = X_1 + X_2 + X_3 + X_4 + X_5 + X_6 + X_7 + X_8$$

By definition, such a deterministic data-generating process has the following features:

- The intercept equals zero.
- The values of each of the eight slope parameters equal unity
- The coefficient of determination (R-squared) equals unity.

Now imagine that we are not interested in building the full model of Y against all its eight drivers. In reality, we usually do not even know how many drivers a given dependent variable has. Instead of building the full model of Y, we are interested in quantifying the strength and direction of the empirical relationship between Y and only one of its assumed drivers, for example X_5. As we will see, even though each of these eight variables constitutes one of the deterministic causes of Y, in the simple regressions of Y against its individual drivers, these explanatory variables either are statistically insignificant or have heavily biased estimates of slope parameters.

Table 6.7 Results of regressing Y against its individual causes (on the basis of the data presented in Table 6.6)

Specification of a regression	Intercept[a]	Slope parameter[a]	R-squared	F statistic
Y_t against X_1	30,95 (10,656)	2,51 (3,875)	0,455	15,01
Y_t against X_2	32,18 (9,243)	1,86 (2,802)	0,304	7,85
Y_t against X_3	29,96 (9,135)	2,06 (3,689)	0,431	13,61
Y_t against X_4	39,59 (7,833)	0,32 (0,317)	0,006	0,10
Y_t against X_5	25,27 (4,990)	2,85 (3,242)	0,369	10,51
Y_t against X_6	28,90 (4,332)	1,91 (1,880)	0,164	3,53
Y_t against X_7	32,76 (11,037)	2,11 (3,160)	0,357	9,99
Y_t against X_8	35,90 (5,545)	0,83 (0,831)	0,037	0,69

Critical value of t-Statistic (at 5% significance level): 2,101
Critical value of F statistic (at 5% significance level): 4,41

Source: Authors
[a]Empirical t-Statistics in brackets

After regressing Y against its individual explanatory variables, we obtain the results shown in Table 6.7.

As can be seen, five out of these eight simple regressions are statistically significant at the 5% significance level and have statistically significant explanatory variables. In the remaining three regressions, both the F statistic as well as the t-Statistic of the slope parameter are below their respective critical values. Furthermore, we note that in all eight regressions, the values of slope parameters are heavily distorted and deviate strongly from their true values of unity. In four of these regressions, the slope parameters are more than twice larger than their true values! Moreover, the intercepts are statistically significant in all eight regressions, while we know that there was no intercept in our data-generating process! So, in such a case of a multivariable relationship between dependent variable and its causes, the results of simple regressions of the dependent variable against its individual drivers may have nothing to do with reality. Such extremely reduced regressions may be extremely distorted and completely unable to quantify the true relationships occurring within multivariable data-generating processes.

If we regress the same dependent variable against more than one but not all of its eight drivers, we obtain yet different results. If we hypothesize that Y has four potential explanatory variables: X_1, X_2, X_3, and X_5, and apply the "general to specific modeling" procedure to select the final set of the explanatory variables, then we begin with the general model with all these four candidate explanatory variables. The results of this general model are presented in Table 6.8.

Table 6.8 Results of regressing Y against X_1, X_2, X_3, and X_5 (on the basis of the data presented in Table 6.6)

Variable	Coefficient	t-Statistic
Intercept	15,19	5,759
X_1	1,54	4,305
X_2	1,27	4,124
X_3	1,06	3,001
X_5	1,41	2,801

Critical value of t-Statistic (at 5% significance level): 2,131
R-squared: 0,891
F statistic: 30,73
Critical value of F statistic (at 5% significance level): 3,06

Source: Authors

As we can see, the estimated model is statistically significant, given its F statistic. Also, all its four explanatory variables are statistically significant. Moreover, the regression is characterized by a relatively large value of R-squared. This model seems to be quite good. However, we again note that all its four slope parameters are biased upward as compared to their true values of unity. The estimated parameters of X_1 and X_5 overstate their true values by as much as 40–50%. Also the intercept is statistically significant, while there was no intercept in our data-generating process.

Finally, we check what happens if we regress the dependent variable Y against all its explanatory variables except for X_8. The results of this regression are presented in Table 6.9. The estimated model is again statistically significant and has seven explanatory variables which are all statistically significant. Moreover, the regression is characterized by relatively high R-squared. Although now the slope parameters of some explanatory variables are close to their true values of unity (e.g., X_1, X_2, or X_6), the other slope parameters are still heavily distorted, particularly X_7. Again the intercept is statistically significant, which is contrary to the true data-generating process. Even in the model with as many as seven out of the eight true drivers of the dependent variable, the obtained regression parameters may be significantly biased.

To sum up, building and interpreting incomplete econometric models is very problematic. Regressions in which not all the relevant explanatory variables are included tend to have significantly biased coefficients, including both slope parameters as well as intercept. Unfortunately, such reduced models are commonly applied in economic research and business applications.

For example, so-called Beta coefficients of stocks, which are used as the proxies for investment risk of individual stocks or portfolios, are often estimated as the slope coefficients of regressions in which the stock returns of a given stock are regressed against stock returns of a broad and diversified portfolio (e.g., some stock index). Such applications of regression analysis are burdened by high risk of obtaining heavily distorted estimates. Regrettably, most economic variables are characterized by rather complex multivariable data-generating processes and it is next to impossible to begin their regression analyses by providing the full lists of candidate explanatory variables. As a result, almost any regression analysis in

Table 6.9 Results of regressing Y against all eight explanatory variables except for X_8 (on the basis of the data presented in Table 6.6)

Variable	Coefficient	t-Statistic
Intercept	6,44	2,195
X_1	0,96	3,488
X_2	1,05	4,793
X_3	1,18	4,558
X_4	0,82	2,966
X_5	1,21	3,318
X_6	0,96	2,890
X_7	0,67	2,767

Critical value of t-Statistic (at 5% significance level): 2,179
R-squared: 0,960
F statistic: 41,02
Critical value of F statistic (at 5% significance level): 2,91

Source: Authors

business and economic research bears the risk of obtaining biased results resulting from estimating incomplete models. In our opinion, this is one of the most important limitations and dangers of using econometric models in business and economics.

6.4   Spurious Regressions and Long-term Trends

Spurious regression is a common problem in many multivariate analyses of time-series data. In most cases the spurious regressions and spurious correlations are brought about by the long-term trends which dominate the short-term changes of the individual investigated variables (Granger and Newbold, 1986).

Suppose that we have time-series data about two related variables:

- The supposed dependent variable, which is the monthly consumption expenditure of an individual consumer in thousands of USD (denoted as Y_t).
- The supposed explanatory variable, which is the monthly salary of the same consumer in thousands of USD (denoted as X_t).

The monthly salary is assumed to be an explanatory variable for the monthly consumption, because of the logical causal relationship between the two. Suppose that we have a time series of these two variables from 15 consecutive months. These data are presented in Table 6.10.

Figure 6.1 shows the time-series data for both variables. In looking at these data, try to answer the following question: is the relationship between the consumption expenditures in a period and the consumer salary in the same period positive or negative? On first sight, it seems to be positive, because at the beginning of the period the values of both variables were generally small as compared to their respective values at the end of the period. Throughout the whole 15-month period under investigation, both variables were showing evident increasing trends,

Table 6.10 The monthly consumption expenditures of an individual consumer in thousands of USD (denoted as Y_t) and the monthly salary of the same consumer in thousands of USD (denoted as X_t)

Number of month	Y^*	X_1
1	4	5
2	2	11
3	10	10
4	8	19
5	15	15
6	11	22
7	22	20
8	16	31
9	24	28
10	22	34
11	31	33
12	27	40
13	38	39
14	34	46
15	40	45

Source: Authors

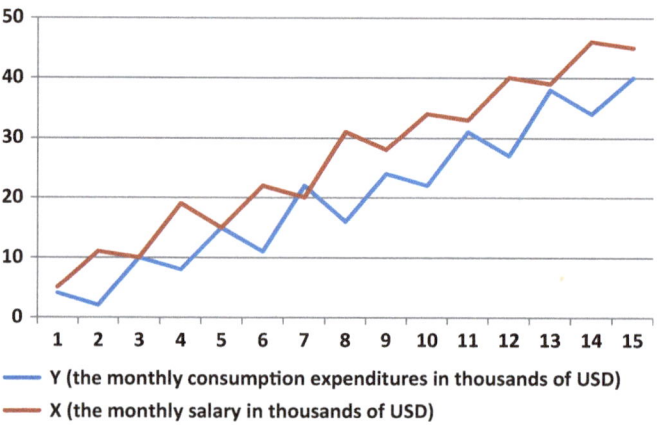

Fig. 6.1 The dependent variable (denoted as Y_t) and the explanatory variable (denoted as X_t), on the basis of the data from Table 6.10. Source: authors

suggesting that fast rising salary was driving up the consumption expenditures. This apparently positive relationship is also confirmed by the simple regression of Y_t against X_t, shown in Fig. 6.2. The slope of that regression (0,83) suggests that the increase of a monthly salary by one dollar tends to boost the monthly consumption expenditures by 83 cents, on average. This allegedly positive relationship seems also to be quite strong, given its R-squared of 0,81.

The simple regression shown in Fig. 6.2 as well as the data presented in Fig. 6.1 suggest that we should answer the question above by stating that the relationship between the consumption expenditures in a period and the consumer salary in the

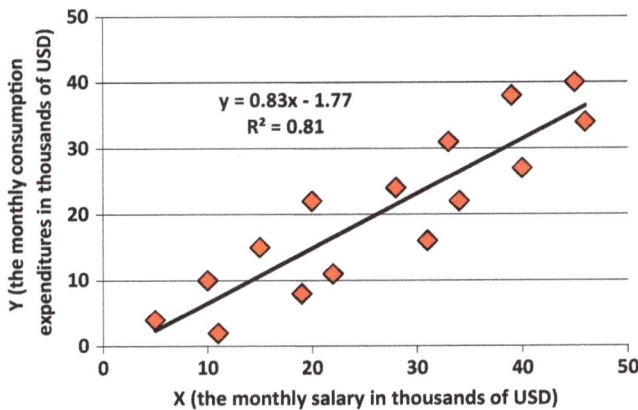

Fig. 6.2 The relationship between Y_t and X_t, on the basis of the data from Table 6.10. Source: authors

same period is positive. However, if the relationship between the dependent variable and the explanatory variable is positive, then how should the monthly consumption change in the following 16th month period if the monthly salary in the following period increases by, say, five thousand USD, as compared to the last month in the sample?

To answer this question we should look at the month-to-month changes of both variables. If we turn our attention back to Table 6.10 or Fig. 6.1, we notice that in periods when the monthly salary rises, the monthly consumption expenditures tend to *decline*! And the months when the monthly salary falls (against previous month), the monthly consumption expenditures usually grow! As the Table 6.10 shows, in the second month of the sample X_t grew while Y_t fell, in the third month X_t decreased while Y_t increased, and so on. This suggests that in the case of the short-term changes (month-to-month), both variables are correlated negatively, despite their positive long-term relationship, presented in Fig. 6.2. This is confirmed by Table 6.11 and Fig. 6.3.

Figure 6.2 showed that the statistical relationship between *levels* of both variables was strongly positive, with a positive slope of 0,83 and R-squared of 0,81. However, Fig. 6.3 informs us that the relationship between *monthly changes* of both variables is even stronger, with R-squared of 0,89, but is now strongly negative, with slope of $-1,19$. The negative slope of that regression suggests that an increase of a monthly salary by one dollar tends to decrease the monthly consumption expenditures by 1,19 dollars, on average. Thus, if in the following 16th month the monthly salary is expected to increase by five thousand USD, we should rather expect the consumption expenditures in that period to drop by approximately six thousand USD ($-1,19 \times 5.000$ USD).

What is behind such contradictory results? The reason is that the relationship quantified on the basis of the data expressed in levels (as shown on Fig. 6.2) missed the fact that the full data-generating process of Y_t is composed of two separate

Table 6.11 The monthly changes of the consumption expenditures (denoted as Y_t) and the monthly changes of the salary (denoted as X_t), on the basis of data from Table 6.10	Number of month	$Y*$	X_1
	1	–	–
	2	−2	6
	3	8	−1
	4	−2	9
	5	7	−4
	6	−4	7
	7	11	−2
	8	−6	11
	9	8	−3
	10	−2	6
	11	9	−1
	12	−4	7
	13	11	−1
	14	−4	7
	15	6	−1

Source: Authors

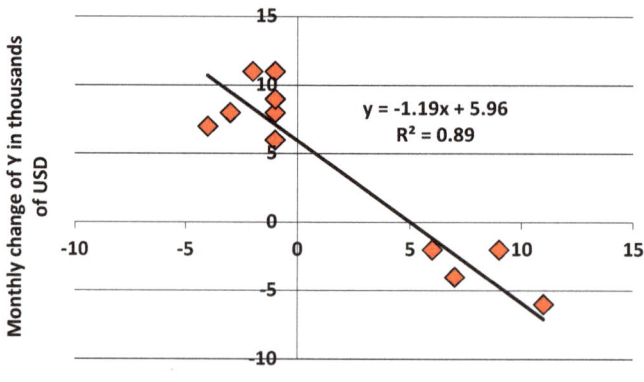

Fig. 6.3 The relationship between monthly changes of Y_t and monthly changes of X_t, on the basis of the data from Table 6.11. Source: authors

factors: strong long-term trend, which is also visible in X_t, and short-term, month-to-month variations around this trend. Furthermore, the long-term rising trend is so strong, as compared to the magnitude of the monthly changes, that it overwhelms the short-term relationship between Y_t and X_t.

Table 6.12 presents the regression of Y_t against X_t, while Table 6.13 contains the results of the regression of monthly changes of Y_t against monthly changes of X_t. For comparative purposes, Table 6.14 shows the results of the regression of levels of Y_t against levels of X_t and a trend variable (numbers of months in the sample).

As Table 6.12 shows, the positive slope parameter of the regression of levels of Y_t against levels of X_t is statistically significant. The whole regression is statistically

Table 6.12 Results of regressing levels of Y_t against levels of X_t (on the basis of the data presented in Table 6.10)

Variable	Coefficient	t-Statistic
Intercept	−1,77	−0,535
X_t	0,83	7,367

Critical value of t-Statistic (at 5% significance level): 2,160
R-squared: 0,807
F statistic: 54,28
Critical value of F statistic (at 5% significance level): 4,67

Source: Authors

Table 6.13 Results of monthly changes of Y_t against monthly changes of X_t (on the basis of the data presented in Table 6.11)

Variable	Coefficient	t-Statistic
Intercept	5,96	8,507
Monthly change of X_t	−1,19	−9,640

Critical value of t-Statistic (at 5% significance level): 2,179
R-squared: 0,886
F statistic: 92,93
Critical value of F statistic (at 5% significance level): 4,75

Source: Authors

Table 6.14 Results of regressing levels of Y_t against levels of X_t and a Trend Variable[a] (on the basis of the data presented in Table 6.10)

Variable	Coefficient	t-Statistic
Intercept	−3,31	3,311
X_t	−1,12	−7,089
Trend Variable[a]	5,84	12,588

Critical value of t-Statistic (at 5% significance level): 2,179
R-squared: 0,986
F statistic: 435,10
Critical value of F statistic (at 5% significance level): 3,89

Source: Authors
[a]The observations of the Trend Variable are the numbers of months in the sample (1, 2, …, 15)

significant as well. However, this regression is unable to capture the short-term relationship between both variables. We already know that month-to-month changes of these variables are strongly negatively related, which is completely missed by the positive value of the slope parameter of 0,83. Such models are called "spurious regressions" because of their doubtful parameters. In time-series models, such a "spuriousness" is usually caused by strong long-term trends of the variables included in the model.

Our hypothetical data for Y_t and X_t were deliberately designed to illustrate the situation in which a strongly negative short-term relationship is associated with a positive slope parameter of the regression model. From the economic point of view, this is not as illogical as it might seem. If we think about the young and hard-working "white-collar," we can imagine a person who in some months is very busy and overloaded with work. In these months, he or she can just have too little spare time to enjoy, with resulting relatively low consumption expenditures in these very months, but such months tend to bring relatively high salaries.

In contrast, less hectic periods afford spending more time on skiing, shopping, traveling, going to the cinemas, pubs, and so on. This gives the reason for observing a negative short-term statistical relationship between earnings and spending. However, the reason might also be much more prosaic, because the salaries earned and booked one month are often paid in the following month. If this is the case, then the income becomes disposable in the month following the one when it was earned.

The economically imaginable negative short-term relationship between earnings and spending does not mean that the positive value of the slope parameter in the regression of levels of Y_t against levels of X_t is senseless. Actually, we should expect that in the long run, both variables would be positively related. When comparing two 5-year periods of anyone's life, usually the relatively high-salary period would be associated with the relatively high consumption, as compared to the 5-year period characterized by much lower earnings. Thus, it might happen in reality that the actual relationship between two variables is strongly negative in the short run while being strongly positive in the long run (or the reverse). Often in such cases, the regression between these variables, estimated on the levels of data, is capable of capturing only the long-term relationship, as in the example investigated here.

In many other cases, it might happen that we obtain statistically significant parameters of a regression between two or more completely unrelated variables, if only they show evident long-term trends. For example, if we regress the number of Internet users in Poland in 2000–2013 (which was showing a strong rising trend) against the population of Detroit in the same period (which was steadily declining), we would observe a *statistically significant* negative relationship. However, both variables have no any relationship in reality. Neither the usage of Internet in Poland is driven by the demographic changes in Detroit nor the inhabitants of Detroit change their place of living in reaction to the development of the Internet in Central Europe. However, if we regress the annual changes of the number of Internet users in Poland against the annual changes of the population of Detroit, then it would turn out that this version of the regression is statistically insignificant. It would mean that the allegedly significant regression estimated in levels of variables is just a spurious one (because no such relationship exists in real life).

Therefore, in modeling time-series regressions, it is very important to distinguish between the long-term and short-term relationships. If we are interested in tracking the long-term relationships, then we can use levels of data, but keeping in mind the risk of obtaining spurious regressions. To measure how the variables are interrelated in the short run, we should either:

• Use period-to-period changes of the original observations which are expressed in levels.
• Account for the long-term trends in the regressions estimated on levels of data, for example by including the trend variable into the set of explanatory variables.

Both approaches are illustrated below.

As Table 6.13 shows, the regression of the monthly changes of Y_t against the monthly changes of X_t is statistically significant and has a statistically significant slope parameter. The slope parameter is now negative and does a good job of capturing the short-term relationship between both variables. However, this regression does not tell us anything about their long-term relationship.

The alternative is to reestimate the regression based on levels of variables, after adding a Trend Variable as an additional explanatory variable. The observations of this Trend Variable are just the numbers of individual periods in the sample (1, 2, ..., 15), if the trend of the dependent variable is linear (which seems to be the case here, when looking at Fig. 6.1). If the trend of the dependent variable is nonlinear, then some other form of the Trend Variable should be considered, for instance the logarithms of the numbers of periods. The results of regressing the levels of Y_t against both the levels of X_t as well as the Trend Variable are presented in Table 6.14.

According to the results contained in Table 6.14, now the slope parameter of a level of X_t is negative, even though this is the same variable as in the regression presented in Table 6.12 (where the value of the respective parameter was 0,83)! Moreover, the value of that parameter does not differ much from the slope parameter of the regression based on monthly changes of the variables (as shown in Table 6.13). This is so because the distorting effect of the long-term rising trends of both Y_t and X_t is now captured by the Trend Variable, which is statistically significant, with the highest t-Statistic of all three regression parameters, and has positive slope parameter. It is also worth noting that the regression of levels of Y_t against levels of X_t and a Trend Variable has a value of the empirical F statistic several times higher than the alternative two regressions. It seems, therefore, that the regression shown in Table 6.14 is the most robust of all three models under consideration.

6.5   Extrapolating In-Sample Relationships Too Far into Out-of-Sample Ranges

One of the most dangerous abuses of econometric models is making predictions or simulations based on assumed or expected values of explanatory variables which are reaching far beyond the ranges of those variables observed in the estimations. We illustrate this with simple example.

Imagine that you are a cost accountant in a manufacturing company. Your company is relatively young and produces ceramic tiles. The monthly production volume in the last 25 months, since the company operates, varied seasonally in the range between 23 thousand square meters and 47 thousand square meters, with production peaks in summer months, when demand is seasonally rising. The monthly total production costs in the same period varied in the range between USD 1.280 thousand and USD 2.380 thousand. However, the demand for ceramic tiles not only is seasonal, but it is also highly cyclical: demand increases when the whole economy booms but it plummets when the whole economy slows down.

Recently, it turned out that the prospects for the whole economy deteriorated with the following decline of demand for building materials, including ceramic tiles. However, this economic slowdown is much deeper than was expected several months ago, and as a result it turned out that your company produced many more ceramic tiles in the last couple of months than justified on the grounds of current demand. Therefore, your company faces the problem of oversupply of finished product. This can be very costly and very dangerous for financial liquidity. The managing board is aware that the production volume in the several forthcoming months must be reduced in order to decrease the amount of excess inventories. However, the chief financial officer is concerned about the impact of the production decline on company's financial results.

Generally speaking, total production costs are composed of fixed costs, which in the short run do not change when production volume changes, and variable costs, which tend to change in tune with changing production volume. The presence of fixed costs in the cost breakdown entails significant impact of shifting production volume on total unit production cost: the lower the production, the higher the total costs per one square meter of production. Because of this, the whole management board is aware that the necessary reduction of production volume will entail the deterioration of company earnings. But the question is: by how much the company earnings will decline as a result of decreased production volume?

The current sale price of tiles is USD 70 per square meter. In light of poor current demand, there is no room for increasing this sale price. The company must rather consider the risk of being forced to reduce the sale price. Therefore, the critical question which must be answered is: what is the *expected* total unit production cost at the reduced production levels? The chief financial officer requested you to run the simulation of total unit production cost at two alternative monthly production levels: 5 thousand square meters and 10 thousand square meters, as compared to the range of production levels between 23 and 47 thousands in the last 25 months.

You decided to estimate the regression line in which total monthly production costs constitute the dependent variable and monthly production volume (in square meters) constitutes the explanatory variable. The statistical data concerning these two variables (presented in order of increasing monthly production volume) are shown in Table 6.15.

As those data confirm, monthly production in this 25-month period varied between 23 and 47 thousands of square meters and total monthly production costs varied in the range between USD 1.280 and USD 2.380 thousand. As we can also see, the unit production cost is negatively correlated to the production volume. At the lowest production levels (23–25 thousand square meters), the unit costs are in the range between USD 53,20 and USD 55,65. In contrast, at the highest production levels (45–47 thousands of square meters), the unit costs are in the range between USD 50,64 and USD 51,56. At the average selling price of USD 70 per square meter, the average pretax profit varied between about USD 14 and USD 19 per square meter.

Your task is to simulate the unit production costs at two alternative production levels: 5 thousand square meters and 10 thousand square meters. You can simulate

Table 6.15 Monthly total production costs (in thousands of USD) and monthly production volume (in thousands of square meters of ceramic tiles) in 25 months (data sorted in order of increasing monthly production volume)

Number of observation	Monthly total production costs[a]	Monthly production volume[b]	Unit production cost (in USD)[c]
1	1.280	23	55,65
2	1.320	24	55,00
3	1.330	25	53,20
4	1.418	26	54,54
5	1.444	27	53,48
6	1.470	28	52,50
7	1.551	29	53,48
8	1.581	30	52,70
9	1.642	31	52,97
10	1.674	32	52,31
11	1.750	33	53,03
12	1.760	34	51,76
13	1.810	35	51,71
14	1.890	36	52,50
15	1.894	37	51,19
16	1.950	38	51,32
17	2.020	39	51,79
18	2.036	40	50,90
19	2.080	41	50,73
20	2.164	42	51,52
21	2.180	43	50,70
22	2.270	44	51,59
23	2.320	45	51,56
24	2.370	46	51,52
25	2.380	47	50,64

Source: Authors
[a]In thousands of USD
[b]In thousands of square meters
[c]Monthly total production costs/monthly production volume

the total monthly production costs at these two production levels and then divide these numbers by respective production levels to arrive at the estimated unit production costs. However, before simulating the total production levels, you need to estimate the relationship between production costs and production volumes. This relationship, based on the data presented in Table 6.15, is shown in Fig. 6.4 and its residuals are presented in Fig. 6.5.

As Fig. 6.4 shows, the relationship between production volume and production costs seems to be linear and strong, with an R-squared of 0,997. Also, the residuals from the estimated regression, presented in Fig. 6.5, seem not to show any regular patterns. However, we cannot state with certainty from such visual inspection

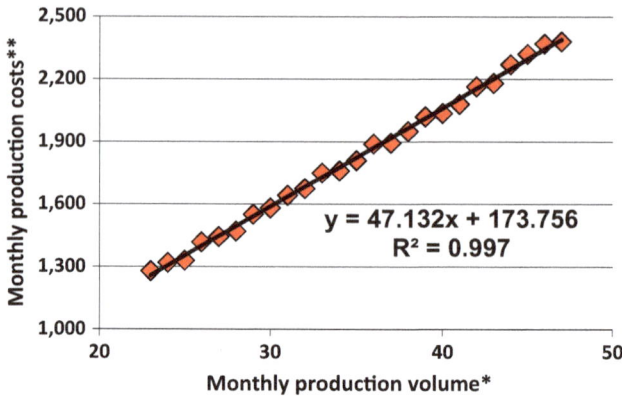

Fig. 6.4 The relationship between monthly total production costs (in thousands of USD) and monthly production volume (in thousands of square meters of ceramic tiles), estimated on the basis of data presented in Table 6.15. *In thousands of square meters. **In thousands of USD. Source: authors

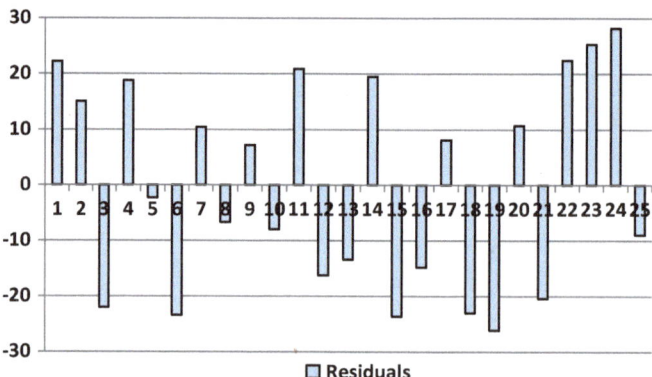

Fig. 6.5 The residuals from the linear regression of monthly production costs against monthly production volume, estimated and presented in Fig. 6.4. Source: authors

whether the assumed linear functional form is correct. We test the validity of this linearity assumption by means of Ramsey's *RESET* test.

To test the correctness of the specification of this linear regression, we need to add some auxiliary explanatory variables constructed as the respective powers of the fitted values from the tested linear regression. We will add two such variables:

- Squared fitted values of the dependent variable obtained from the tested regression (denoted as $FITTED^2$),
- Cubic fitted values of the dependent variable obtained from the tested regression (denoted as $FITTED^3$).

Table 6.16 Auxiliary regression estimated for Ramsey's *RESET* test conducted for the linear regression of monthly production costs against monthly production volume

Variable	Coefficient	*t*-Statistic
Intercept	233,03	0,474
X	48,13	0,858
$FITTED^2$	0,00	−0,116
$FITTED^3$	0,00	0,210
R-squared: 0,998		

Source: Authors

Then we need to reestimate the tested linear regression after adding these two auxiliary explanatory variables. The results of this auxiliary regression are presented in Table 6.16.

As we can see in Fig. 6.4, the *R*-squared of the tested model equals 0,997. Table 6.16, in turn, informs us that the *R*-squared of the auxiliary regression equals 0,998. The tested regression was estimated on the basis of $n = 25$ observations, and it includes one explanatory variable ($k = 1$). Therefore, our empirical *F* statistic is computed as follows:

$$F = \frac{\left(R_2^2 - R_1^2\right)/2}{\left(1 - R_2^2\right)/(n - (k + 3))} = \frac{(0,998 - 0,997)/2}{(1 - 0,998)/(25 - (1 + 3))} = 1,53.$$

Now we need to obtain the critical value of F^*. For $n = 25$ and $k = 1$, we obtain:

$$m_1 = 2,$$

$$m_2 = n - (k + 3) = 25 - (1 + 3) = 21.$$

For the level of statistical significance set at $\alpha = 0,05$, and determined values of m_1 and m_2, the critical value of F^* from the Fisher–Snedecor table equals 3,47.

Finally, we compare the value of the empirical statistic *F* with its critical value F^*. In our case $F < F^*$, which means that H_0 is not rejected, and we conclude that none of the two auxiliary variables is statistically significant. Therefore, we can conclude that the linear regression of monthly production costs against monthly production volume is correctly specified.

To sum up, the Ramsey's *RESET* test corroborated that the assumed linear functional form of the relationship between monthly production costs and monthly production volume is justified. If this model is well specified, we can proceed to simulating the total and unit production costs at monthly production levels set at 5 thousand square meters and 10 thousand square meters.

On the basis of this linear regression (presented in Fig. 6.4), we can simulate the expected monthly total production costs as follows:

- For monthly production equaling 5 thousands square meters:

$$Y = 173,756 + 47,132\,X = 173,756 + 47,132 \times 5 = 409,42.$$

- For monthly production equaling 10 thousands meters:

$$Y = 173,756 + 47,132\,X = 173,756 + 47,132 \times 10 = 645,08.$$

Now we can simulate the expected monthly unit production costs. They equal USD 81,88 per square meter when monthly production is 5 thousand square meters $(409,42/5 = 81,88)$ and USD 64,51 per square meter when monthly production is 10 thousand square meters $(645,08/10 = 64,51)$. If the selling price may be maintained at USD 70 per square meter, then we can expect that the company will incur a unit loss of USD 11,88 when production level is 5 thousands square meters $(70,00 - 81,88 = -11,88)$ and it will still earn a positive unit gross profit of USD 5,49 when production level is 10 thousand square meters $(70,00 - 64,51 = 5,49)$.

Meanwhile, after reducing monthly production to 10 thousands square meters, the company found that the actual unit production cost amounted to USD 101,90 (instead of expected USD 64,51)! And after reducing monthly production level further to 5 thousand square meters, the actual unit production cost jumped to USD 190,00 (instead of expected USD 81,88)! So, at monthly production set at 10 thousand meters the company incurred a unit loss of USD 31,90 $(70,00 - 101,90 = -31,90)$ and at monthly production set at 5 thousands meters the company incurred a unit loss of USD 120,00 $(70,00 - 190,00 = -120,00)$.

What could have happened which caused such huge underestimation of unit production costs by our linear regression of monthly production costs against monthly production volume? This linear regression seemed to be robust, well specified and having high goodness of fit, but these conclusions are valid only within the sample on which the regression was estimated.

In our sample of data from 25 months, the monthly production volume varied in the range between 23 thousand square meters and 47 thousand square meters, then we simulated the value of the dependent variable for much lower values of explanatory variable. Meanwhile, the regression statistics within the sample, including conclusions from Ramsey's *RESET* test, may not hold outside this sample. Particularly, we can never be sure that the functional form of the investigated relationship, observed within the range of data which we have at our disposal, will be still valid when we move outside the boundaries of this range. In our case, we extrapolated the estimated linear relationship rather far from the range on which that relationship was quantified.

It often happens that the relationship, which is apparently linear within the sample, turns out to be evidently nonlinear when the range of data is extended. In our case, the true relationship between monthly production costs and monthly production volume, but in the range of monthly production volume between 1 thousand square meters and 52 thousand square meters (instead of the narrower range between 23 thousand square meters and 47 thousand square meters) may look as in Fig. 6.6. This relationship is based on the data from Table 6.17. In this table, all the observations marked by italics (i.e., the observations with monthly production volume between 23 thousand square meters and 47 thousand square meters) are the same as shown in Table 6.15.

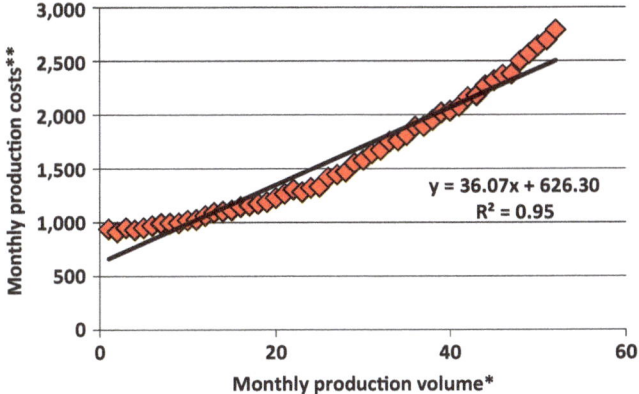

Fig. 6.6 The relationship between monthly total production costs (in thousands of USD) and monthly production volume (in thousands of square meters of ceramic tiles), estimated on the basis of data presented in Table 6.17. *In thousands of square meters. **In thousands of USD. Source: authors

As we can see in Fig. 6.6, now the relationship between production costs and production volume seems nonlinear. Although within the range of production volume between 23 and 47 thousand square meters, the production costs seemed to decline linearly with decreasing production volume; at the lower levels of production (below 20 thousand square meters), the costs are falling much slower with falling production volume. This caused our linear regression to heavily understate the true production costs at exceptionally low levels of production. It seems evident that the relationship between these two variables should be better captured by a power or exponential function, instead of a linear function. However, we are not usually able to test the true functional form within the wide range of data if we do not have a statistical sample with such wide dispersion and instead we have limited dataset with smaller dispersion.

As the above example showed, it is very risky to use regression model to produce forecasts or simulations of a dependent variable when the values of explanatory variables are lying outside their respective ranges observed in the sample on which the model was estimated. We must always be aware that even the relationships which seem to have obvious functional forms in some samples may turn out to show completely different patterns in wider ranges of data.

6.6 Estimating Regressions on Too Narrow Ranges of Data

In the preceding section, we discussed the problem of non-linearities occurring outside the range of the sample on which the linear regression is built. In that case, the relationship between variables which seemed to be linear within the sample turned out to be nonlinear when the ranges of data were extended. As we remember,

Table 6.17 Monthly total production costs and monthly production volume in wider range of production volume (original data from Table 6.15 are marked by italics)

Monthly total production costs[a]	Monthly production volume[b]	Monthly total production costs[a]	Monthly production volume[b]
940	1	1.444	27
910	2	1.470	28
950	3	1.551	29
930	4	1.581	30
950	5	1.642	31
970	6	1.674	32
990	7	1.750	33
992	8	1.760	34
995	9	1.810	35
1.019	10	1.890	36
1.025	11	1.894	37
1.062	12	1.950	38
1.091	13	2.020	39
1.101	14	2.036	40
1.112	15	2.080	41
1.144	16	2.164	42
1.158	17	2.180	43
1.173	18	2.270	44
1.189	19	2.320	45
1.226	20	2.370	46
1.265	21	2.380	47
1.305	22	2.500	48
1.280	23	2.567	49
1.320	24	2.641	50
1.330	25	2.695	51
1.418	26	2.790	52

Source: Authors
[a]In thousands of USD
[b]In thousands of square meters

the extrapolation of the allegedly linear in-sample relationship outside the sample resulted in dramatically inaccurate forecast of the dependent variable.

In the example presented in Sect. 6.5, the statistical relationship which appeared to be nonlinear in the full range of data seemed to be linear when we had only reduced datasets. However, this relationship still existed and was very strong (having R-squared of 0,997) in the reduced sample. Therefore, even though the hidden nonlinearity (unobservable in the reduced sample) resulted in inaccurate out-of-sample predictions obtained from the linear regression, the linear regression estimated on the reduced dataset correctly suggested the true direction and strength of the modeled relationship.

In such a case, there is no significant risk of stating that the relationship does not exist on the grounds of the reduced sample, while it really exists in the full dataset.

However, there is another potential pitfall related to estimating regressions on the basis of reduced datasets. This is related to the risk of stating that the statistical relationship between two or more variables is statistically insignificant, while in reality it is significant and relatively strong within the full range of variability of variables included in the model.

Suppose you are a financial analyst at some industrial company. The Chief Financial Officer (your boss) has recently suggested at the meeting of the Managing Board that your company is burdened by the relatively high cost of debt because of the interest rate paid to banks due to loans. The CFO has suggested that the cost of debt incurred by competitors is varying significantly in relation to the significant differences between profitability of individual companies, since the less profitable companies are deemed to be more risky borrowers and as a result pay higher interest rates. If this is the case, then boosting your company's operating profitability would lower the financial burden brought about by having to pay relatively high interest rates on debt. To test this hypothesis about the negative relationship between profitability and cost of debt, your CFO has requested you to collect the respective data of your competitors and estimate the regression line between these two variables.

There are 100 companies operating in this industry. However, only 30 out of 100 are public (listed on the stock exchange) and the remaining 70 are private. This is important because you can collect detailed data, contained in full financial statements, of public companies, which are obligated to report their financial results on a regular basis. In contrast, the private companies do not publish any detailed and updated information about their financial situation.

After collecting the annual reports of all these 30 competitors, you estimated the simple linear regression in which the cost of debt (expressed as interest rate) is the dependent variable and operating profitability (annual operating profit divided by annual sales revenues) is the only explanatory variable. This relationship within the sample of 30 public companies is illustrated in Fig. 6.7. Clearly, the cost of debt and operating profitability seem not to be statistically related to each other.

We can see that the operating profitability within this sample of 30 public companies varies between -1% and 6%. The majority of companies are profitable, but there are two which incur operating losses. Similarly, the interest rates on debt vary in the range between about 6,5% and slightly more than 9,5%. We can conclude, therefore, that the analyzed companies are very diverse in terms of their profitability and costs of debt. However, the statistical relationship between the two variables is insignificant in this sample, even without applying any formal tests for its significance. Although the slope parameter is negative, the regression line seems to be almost parallel to the horizontal axis and its R-squared equals only 0,001.

Given that the actual interest rates on debt vary in a relatively wide range in this sample, it seems that these intercompany differences in the cost of debt are driven by factors other than operating profitability. Among these factors may be total indebtedness, variability of profits, age, and size of the company. So you could conclude, on the grounds of these data, that there is not any statistically significant relationship between cost of debt and company profitability, thus rejecting the CFO's hypothesis.

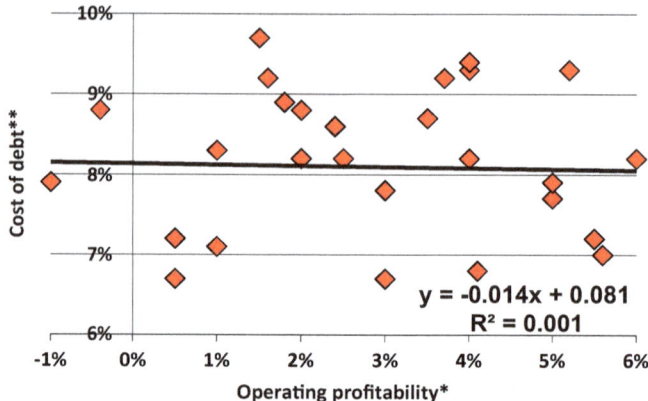

Fig. 6.7 The relationship between cost of debt and operating profitability of 30 public companies from Your industry. * Annual operating profit/annual sales revenues. ** The average interest rate paid by a given company to the banks. Source: authors

However, there are 100 companies, not just 30, operating in this industry. It could happen that the inclusion of the remaining 70 companies into the sample, would change the findings. Suppose that most of these private companies are characterized by either below-average or above-average profitability as compared to the public companies. As a result, the profitability in the whole sample of 100 companies varies between −10% and 10%, instead of between −1% and 6% in the subsample of 30 public companies. If this is the case and if the operating profitability is one of the drivers of cost of debt, then we could obtain the relationship presented in Fig. 6.8.

Is it probable that private companies could differ so much in terms of operating profitability from public ones? Although the differences of such extent as in this example are rather theoretical, it may happen that:

- Those companies from the industry, which are characterized by negative profitability, are not listed on the stock exchange, because it would be difficult for them to attract stock investors, while there are some banks willing to lend them money at much higher cost of debt.
- Those companies from the industry, which are characterized by above-average positive profitability, are not listed on the stock exchange, because they are so profitable that they do not need any equity capital from stock market investors. Instead, they are self-sufficient in providing the capital for further growth from current profits. Alternatively, their owners might be motivated to keep these companies private in order not to disclose this very information about above-average profitability to the competitors.

Anyway, it may happen that the subsample of public companies has a completely different distribution of profitability than the subsample of private

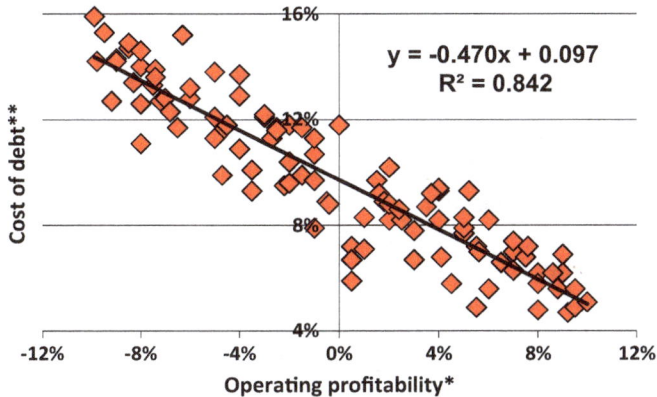

Fig. 6.8 The relationship between cost of debt and operating profitability of 100 companies from your industry (including 30 public and 70 private companies). *Annual operating profit/annual sales revenues. **The average interest rate paid by a given company to the banks. Source: authors

companies. It could also mean that the relationship between cost of debt and operating profitability, observed in the whole population of companies within this industry, is much stronger than in the subsample of public companies. In Fig. 6.8, the subsample of public companies (the same observations which were shown in Fig. 6.7) is included in the whole sample. It is now evident that the relationship between the two investigated variables is strong (given its R-squared of 0,842) and having the expected direction, manifested in the slope parameter equaling −0,470.

As can be seen in Fig. 6.8, companies with low profitability (say, between −10% and 0%) are featured by a relatively high cost of debt, in the range between 8% and 16%, while companies with higher profitability, between 0% and 10%, are burdened by a relatively low cost of debt, in the range between about 5% to about 10%. We can conclude, therefore, that operating profitability is a strong driver of cost of debt within this particular industry. The CFO was, therefore, right: boosting your company's operating profitability will probably help in lowering the interest rates on your company's debt, provided that this increase of profitability is of sufficient magnitude. The relationship which was not visible in the reduced sample of 30 public companies is much more evident if the ranges of data of both variables are extended.

This was another example confirming that applying statistical tools in a mechanical way can be dangerous. Here, basing the regression line on reduced dataset resulted in obtaining incorrect findings, according to which there is no statistically significant relationship between the two variables. As we saw, arriving at correct findings required using more complete datasets, many times unavailable for the analyst. The question, therefore, arises: how can we know that the datasets which we have at our disposal have the ranges wide enough to ensure obtaining correct and credible inferences about the investigated relationships? Unfortunately, there is no statistical tool which could be objectively applied for this purpose.

Sometimes analysts apply some "rules of thumb" related to the coefficient of variation in checking whether their data have sufficient variability, such as assuming that the coefficient of variation of any variable should exceed 25%. However, the power of such "rules of thumb" is unknown and probably weak. The credibility of any statistical analysis is ensured only when application of statistical tools is combined with using common sense and professional experience in the field. Here, a financial analyst experienced in evaluating the financial situation of companies in that particular industry would probably know that there are some companies with operating profitability reaching outside the range observed in the subsample of 30 public companies. Knowing this we should strive to include these observations in our dataset, controlling for any outliers. If, however, we cannot extend our sample to get the wider ranges of data, then it is often safer to abandon using the statistical tools at all or at least to keeping in mind that our findings are valid only for these narrower data ranges. Lack of statistical findings is better and safer than having incorrect statistical findings.

6.7 Ignoring Structural Changes Within Modeled Relationships and Within Individual Variables

6.7.1 Introduction

In this section, we discuss the pitfalls of neglecting structural changes, also called structural breaks, occurring within the relationships modeled by regressions. We will touch two types of structural changes:

- Structural changes occurring when the nature of variables is stable, but the relationships between variables in the model are shifting.
- Structural changes occurring when the relationships between variables in the model are stable but the nature and structure of individual variables are shifting.

The first of these two circumstances has already been discussed in Chap. 5, where we presented the example of the gradually changing structure of relationships between the dependent variable and explanatory variables. We also presented the method of testing and dealing with structural breaks by means of dummy variables, and we also discussed the pitfalls of interpreting regression parameters in case of structural breaks. However, in this section, we extend this discussion and we illustrate the impact of structural changes on the properties of regression residuals.

The second situation, that of stable relationships between modeled variables but changing structure and nature of these variables, is much more dangerous and, unfortunately, more difficult to detect and deal with. Moreover, the situation frequently occurs in many economic processes, although often analysts and statisticians are not aware of being trapped by such structural breaks. As usually, we will show the numerical example of such structural breaks.

6.7.2 Structural Changes in Relationships Between Variables

Table 5.17 in Chap. 5 presented hypothetical data about monthly total production costs (in thousands of USD) and monthly production volume (in physical units, e.g., kilograms) of a company from 24 consecutive months. Our analytical task was to separate, on the basis of these data, the fixed and variable elements in cost structure, which are not directly observable.

In that example, in the first month of the sample, the fixed costs amounted to USD 1.000 thousand while the variable cost per unit of production amounted to USD 2 thousand. Both these numbers are unobservable and hidden in the total monthly production costs. In contrast, in the last month, the fixed costs amounted to USD 2.750 thousand while the variable cost per unit of production amounted USD 1,31 thousand. The structure of these production costs (i.e., fixed costs vs. variable costs) was constantly changing within the analyzed sample, with gradually rising volume of monthly fixed expenses and declining unit variable costs. This could be caused by some variable costs becoming less and less elastic against the changes of monthly production.

Our intention was to split, on the grounds of these 24-month data, the total costs into fixed costs and variable costs per unit. We did this by estimating the following simple linear regression:

$$TC_t = \alpha_0 + \alpha_1 TP_t,$$

where:

TC_t—total production costs in period t,

TP_t—total production volume (physical units or other cost drivers) in period t,

α_0—intercept interpreted as average periodic fixed costs (i.e., costs which do not change with changing production volume),

α_1—slope parameter interpreted as average variable production costs per unit of production (i.e., increase/decrease of total production costs caused by increase/decrease of production volume by one unit).

After running this regression, we obtained the parameters presented in Table 6.18.

According to the numbers presented in Table 6.18, the estimated monthly fixed costs in the investigated period amounted to USD 1.933 thousand while estimated unit production costs equaled USD 1,52 thousand. But these parameters do not reflect any economically relevant numbers, from the point of view of decision-making within the company. For example, the obtained estimate of monthly fixed

Table 6.18 Results of linear regression of monthly total production costs (in thousands of USD and denoted as TC_t) against monthly production volume (denoted as TP_t), estimated on the basis of data from the last two columns of Table 5.17

Variable	Coefficient	t-Statistic
Intercept	1.932,72	8,474
TP_t	1,52	9,310

Critical value of t-Statistic (at 5% significance level): 2,074
R-squared: 0,798
F statistic: 86,68
Critical value of F statistic (at 5% significance level): 4,30

Source: Authors

costs (1.933) is misleading because it does not correspond to true costs at the extreme observations nor average fixed costs in the whole 24-month period. If we again look at the data in Table 5.17, we can see that:

- The actual fixed costs at the beginning of the sample equaled USD 1.000 thousand monthly, which is about 48,3% less than the estimate from the regression.
- The actual fixed costs at the end of the sample (which are by far most important for management decisions) equaled USD 2.750 thousand monthly, which is about 42,3% more than the estimate from the regression.
- The arithmetic average of actual monthly fixed costs in the whole 24-month period equals USD 1.737,50 thousand monthly, which is more than 10% less than the estimate from the regression,
- The median of actual monthly fixed costs in the whole 24-month period equals USD 1.600 thousand monthly, which is more than 17% less than the estimate from the regression.

As we can see, the estimated fixed costs (USD 1.933 thousand per month) are rather not useful for making any economic decisions in this company. Particularly dangerous would be to treat this number as the estimate of current (i.e., at the end of the 24-month period) volume of monthly fixed costs, because this estimate heavily understates the true current fixed expenses (which equal USD 2.750 thousand monthly). If this regression-based estimate of fixed costs corresponds to any true costs, these are the costs from the 16th month (outdated by some eight months). Basing economic decisions (e.g., pricing products or evaluating business risks) on such heavily distorted estimates of fixed costs is clearly not a good idea.

The same qualifications apply to the obtained estimate of unit variable costs. This number (equaling USD 1,52 thousand per unit of production) understates the actual variable costs at the beginning of the sample (2,00), as well as the average and median variable costs in the whole sample (1,66), but it heavily overstates the actual recent variable costs (1,31).

To summarize, neglecting structural changes occurring within the modeled relationships may result in obtaining allegedly good and credible regressions (having a high R-squared, F statistics, and t-Statistics), which, however, produce misleading conclusions about the nature of analyzed processes. In the above example, the obtained regression suggested (incorrectly) that the structure of company's production costs is relatively flexible while in reality it is much more fixed.

Fortunately, such complete neglect for structural breaks usually results in obtaining regression residuals which immediately signal that something is wrong when verifying the model (because these residuals tend to be autocorrelated or heteroscedastic). Table 6.19 presents the underlying data as well as the fitted values and residuals from the regression shown in Table 6.18.

Even the first glimpse at the obtained residuals immediately suggests that they are not "noisy." The first part of the sample (up to the 14th observation) is dominated by negative residuals while the remaining part (from the 15th observation onward) consists of only positive residuals. The results of the two tests presented in Table 6.20 confirm that the regression shown in Table 6.18 is evidently incorrect. The residuals from estimated simple regression are both serially autocorrelated and heteroscedastic.

Adding a dummy variable to account for structural change to the regression effectively eliminates the autocorrelation and heteroscedasticity of residuals. Table 6.21 presents the underlying data as well as the fitted values and residuals from the regression of monthly costs with a dummy variable for structural break. Table 6.22 in turn shows the results of autocorrelation and heteroscedasticity of residuals presented in the last column of Table 6.21. As can be seen, after including the dummy variable for structural change into the regression of monthly production costs against monthly production volume, the obtained residuals are much more "chaotic." They do not present any statistically significant autocorrelation or heteroscedasticity.

Unfortunately, however, dealing with the structural breaks by dummy variables does not guarantee that the model parameters will no longer be distorted. As was already discussed in Chap. 5, although inclusion of a dummy variable for structural change resulted in significant improvement of goodness of fit, the extended model, including a dummy variable for structural break, produced an even more heavily biased estimate of current unit variable costs than a simple regression without any dummy variables.

To sum up, if we have any statistical, theoretical, or experience-based reason to suppose that the structure of modeled relationship is shifting, we should always be very careful in interpreting the individual regression parameters.

6.7.3 Structural Changes Inside Individual Variables in the Model

The problem of the changing structure of modeled relationships, although very common and important, is usually indirectly signaled by one or more of the tests

Table 6.19 Observations, fitted values, and residuals of the regression presented in Table 6.18

Number of period	Monthly production volume in physical units (explanatory variable)	Total monthly production costs[a] (dependent variable)	Fitted values of dependent variable	Residuals
1	1.000	3.000	3.452,93	−452,93
2	1.500	4.005	4.213,03	−208,03
3	2.000	4.980	4.973,14	6,86
4	1.800	4.588	4.669,09	−81,09
5	1.200	3.456	3.756,97	−300,97
6	1.500	4.025	4.213,03	−188,03
7	1.400	3.848	4.061,01	−213,01
8	1.000	3.140	3.452,93	−312,93
9	800	2.808	3.148,89	−340,89
10	1.100	3.353	3.604,95	−251,95
11	1.400	3.880	4.061,01	−181,01
12	1.200	3.554	3.756,97	−202,97
13	1.100	3.454	3.604,95	−150,95
14	1.600	4.326	4.365,05	−39,05
15	2.200	5.326	5.277,18	48,82
16	2.000	5.050	4.973,14	76,86
17	1.700	4.634	4.517,07	116,93
18	1.000	3.640	3.452,93	187,07
19	1.300	4.148	3.908,99	239,01
20	1.100	3.923	3.604,95	318,05
21	800	3.570	3.148,89	421,11
22	1.200	4.194	3.756,97	437,03
23	1.400	4.526	4.061,01	464,99
24	1.000	4.060	3.452,93	607,07

Source: Authors
[a]In thousands of USD

used in verifying the correctness of the model. However, there is another type of structural change, which is more subtle, difficult to detect by means of statistical tools and requiring strong and deep economic knowledge to be correctly dealt with. This is related to structural changes occurring within the individual variables included in the model, instead of structural changes of relationships between the variables.

Suppose that you are a financial analyst and you work for an investment bank or stock brokerage house. Your task is to make a valuation of one public company and to recommend whether to buy or sell shares of this company to the clients of your employer. In this case, you are an outside analyst, not from the analyzed company, so you can base your analysis and recommendation only on publicly available data.

You then decide to base your valuation mainly on the financial and operating data published by the investigated company in its quarterly reports. One of the elements of your analysis is the evaluation of the share of fixed costs in the total

Table 6.20 Tests for autocorrelation and heteroscedasticity of residuals conducted for the regression presented in Table 6.18

F-test for autocorrelation of residuals (of up to fourth order)	
R-squared of regression of residuals against lagged (up to fourth order) residuals	0,951
Empirical F statistic (for $n = 20$ and $k = 4$)	73,06
Critical value of F statistic	3,06
Conclusion:	Residuals are autocorrelated
ARCH-LM test for heteroscedasticity of residuals (of up to fourth order)	
R-squared of regression of squared residuals against lagged (up to fourth order) squared residuals	0,789
Empirical F statistic (for $n = 20$ and $k = 4$)	14,05
Critical value of F statistic	3,06
Conclusion:	Residuals are heteroscedastic

Source: Authors

cost structure of that company. This is one of the important operating risk measures, because the higher the share of fixed costs, the higher the variability of earnings, a phenomenon known as "operating leverage." Suppose that from the company's quarterly reports you can derive the data about total operating costs in any quarter and total production volume in the same quarter. However, these public data are available only on aggregate level, relating to the analyzed company as a whole.

Meanwhile, the company has three manufacturing facilities, located in three different countries, with three different cost structures. For example, a manufacturing facility located in a country with strong employee protection may have much higher fixed salaries and other employee-related expenses, such as social security benefits, while manufacturing operations located in another country, with much looser employee-protection regulations, may have much more flexible cost structure (with mostly variable costs). Table 6.23 presents hypothetical data about quarterly operating costs and quarterly production volumes from 24 quarters in three production facilities (data from company's internal reports, not available for outside analysts) as well as the aggregated data (which are reported publicly and constitute the sums of data from three production facilities).

The third row from the top of Table 6.23 provides actual quarterly fixed operating costs and actual variable unit operating costs for three individual manufacturing factories. According to these data, Factory A is characterized by highest fixed costs (equaling USD 2.000 thousand per quarter) and relatively low variable costs (equaling USD 0,50 thousand per one unit of production). In contrast, Factory C is featured by relatively low fixed costs (only USD 500 thousand per quarter) compensated by relatively high variable costs (USD 2 thousand per unit). It suggests that these three production facilities, although producing identical products, are very diverse in terms of their operating cost structures.

Table 6.21 Observations, fitted values, and residuals of the regression of monthly production costs with a dummy variable for structural break

Number of period	Monthly production volume in physical units (explanatory variable)	Dummy variable for structural break (explanatory variable)	Total monthly production costs[a] (dependent variable)	Fitted values of dependent variable	Residuals
1	1.000	1	3.000	3.159,32	−159,32
2	1.500	1	4.005	4.073,89	−68,89
3	2.000	1	4.980	4.988,46	−8,46
4	1.800	1	4.588	4.622,63	−34,63
5	1.200	1	3.456	3.525,15	−69,15
6	1.500	1	4.025	4.073,89	−48,89
7	1.400	1	3.848	3.890,98	−42,98
8	1.000	1	3.140	3.159,32	−19,32
9	800	1	2.808	2.793,49	14,51
10	1.100	1	3.353	3.342,23	10,77
11	1.400	1	3.880	3.890,98	−10,98
12	1.200	1	3.554	3.525,15	28,85
13	1.100	1	3.454	3.342,23	111,77
14	1.600	1	4.326	4.256,81	69,19
15	2.200	1	5.326	5.354,29	−28,29
16	2.000	1	5.050	4.988,46	61,54
17	1.700	1	4.634	4.439,72	194,28
18	1.000	0	3.640	3.799,67	−159,67
19	1.300	0	4.148	4.348,41	−200,41
20	1.100	0	3.923	3.982,58	−59,58
21	800	0	3.570	3.433,84	136,16
22	1.200	0	4.194	4.165,50	28,50
23	1.400	0	4.526	4.531,33	−5,33
24	1.000	0	4.060	3.799,67	260,33

Source: Authors
[a]In thousands of USD

In this example, we assumed that throughout the whole period of 24 quarters, the fixed and variable costs of individual factories were constant, unlike in previous example of structural break, where the cost structure was gradually changing from period to period. For illustrative purposes, we also made a very strong and unrealistic assumption according to which total operating costs of individual factories are deterministically driven by changes in production levels of these factories, without any other factors influencing total operating costs. This means that total quarterly costs can be computed without any error based on quarterly production level, quarterly fixed costs, and unit variable costs. For example, in Factory A in the first quarter the production volume was 1.000 units, fixed costs equaled 2.000

Table 6.22 Tests for autocorrelation and heteroscedasticity of residuals conducted for the regression of monthly production costs with dummy variable for structural break

F-test for autocorrelation of residuals (of up to fourth order)	
R-squared of regression of residuals against lagged (up to fourth order) residuals	0,281
Empirical *F* statistic (for $n = 20$ and $k = 4$)	1,47
Critical value of *F* statistic	3,06
Conclusion:	Residuals are not autocorrelated
ARCH-LM test for heteroscedasticity of residuals (of up to fourth order)	
R-squared of regression of squared residuals against lagged (up to fourth order) squared residuals	0,053
Empirical *F* statistic (for $n = 20$ and $k = 4$)	0,21
Critical value of *F* statistic	3,06
Conclusion:	Residuals are homoscedastic

Source: Authors

thousands of USD, and variable costs equaled 0,50 thousands of USD per unit, which results in total operating costs equaling 2.500 thousands of USD (2.000 + 1.000 × 0,50). Similarly, in Factory B in the last quarter the production volume was 1.200 units, fixed costs equaled 1.000 thousands of USD, and variable costs equaled 1,50 thousands of USD per unit, which results in total operating costs of 2.800 thousands of USD (1.000 + 1.200 × 1,50).

Sums of total operating costs incurred in three individual factories as well as sums of production volumes of these factories give the respective data aggregated for the whole company. These aggregated numbers are reported in company's quarterly reports, on which you as an analyst base your valuation. For instance, the aggregate operating costs in the first quarter amounted to 9.000 thousands of USD (2.500 + 2.800 + 3.700), and the company-level production volume in the same quarter was 3.800 units (1.000 + 1.200 + 1.600). Likewise, the sum of fixed costs of the three factories constitutes the total fixed costs of the whole company, equaling 3.500 thousands of USD (2.000 + 1.000 + 500). However, this is the number which we know on the grounds of these factory-level data, but which you cannot observe directly from the aggregate company-level data, available from the company's quarterly reports.

Instead of knowing that the total company-level fixed costs equal 3.500 thousands of USD quarterly and are driven in a deterministic way by the changes in production levels at three factories, you must estimate these quarterly company-level fixed costs on the basis of the publicly available data from the last two columns of Table 6.23. If you regress the total quarterly operating costs (shown in the next to last column of Table 6.23) against total quarterly production (shown in the last column of Table 6.23), you will obtain the results presented in Fig. 6.9.

Table 6.23 Fixed costs, variable costs, total costs, and production volume for three production facilities (factories) as well as for the company as a whole

Number of quarter	Factory A		Factory B		Factory C		Whole company	
	FOC[a]	VOC[b]	FOC[a]	VOC[b]	FOC[a]	VOC[b]	FOC[a]	VOC[b]
	2000	0,50	1 000	1,50	500	2,00	3 500	–
	TOC[c]	PV[d]	TOC[c]	PV[d]	TOC[c]	PV[d]	TOC[c]	PV[d]
1	2.500	1.000	2.800	1.200	3.700	1.600	9.000	3.800
2	2.600	1.200	2.500	1.000	4.300	1.900	9.400	4.100
3	3.000	2.000	2.050	700	3.300	1.400	8.350	4.100
4	2.800	1.600	2.650	1.100	3.700	1.600	9.150	4.300
5	2.650	1.300	2.950	1.300	2.900	1.200	8.500	3.800
6	2.700	1.400	3.100	1.400	2.500	1.000	8.300	3.800
7	2.500	1.000	2.650	1.100	3.900	1.700	9.050	3.800
8	2.550	1.100	3.250	1.500	4.300	1.900	10.100	4.500
9	2.400	800	4.000	2.000	3.500	1.500	9.900	4.300
10	2.500	1.000	3.550	1.700	3.100	1.300	9.150	4.000
11	2.650	1.300	3.250	1.500	3.700	1.600	9.600	4.400
12	2.600	1.200	2.800	1.200	2.100	800	7.500	3.200
13	2.450	900	2.950	1.300	4.100	1.800	9.500	4.000
14	2.800	1.600	2.650	1.100	3.900	1.700	9.350	4.400
15	2.950	1.900	2.350	900	2.500	1.000	7.800	3.800
16	2.850	1.700	3.100	1.400	3.300	1.400	9.250	4.500
17	2.700	1.400	3.250	1.500	3.100	1.300	9.050	4.200
18	2.650	1.300	2.650	1.100	2.300	900	7.600	3.300
19	2.700	1.400	2.500	1.000	4.100	1.800	9.300	4.200
20	2.600	1.200	3.250	1.500	3.900	1.700	9.750	4.400
21	2.450	900	4.450	2.300	3.300	1.400	10.200	4.600
22	2.600	1.200	3.700	1.800	3.500	1.500	9.800	4.500
23	2.650	1.300	3.550	1.700	2.900	1.200	9.100	4.200
24	2.750	1.500	2.800	1.200	2.500	1.000	8.050	3.700

Source: Authors
[a]FOC = fixed costs per quarter in thousands of USD
[b]VOC = variable costs per unit of production in thousands of USD
[c]TOC = total operating costs in a given quarter in thousands of USD, computed for individual factories as follows: TOC = FOC + (VOX × PV)
[d]PV = production volume in a given quarter in physical units

As can be seen in Fig. 6.9, the statistical relationship between company-level quarterly operating costs and company-level quarterly production volume seems to be quite strong (with R-squared of 0,767), consistent with expectations as regards its direction and having parameters which seem to have reasonable values. However, the estimated company-level quarterly fixed costs equal approximately 1.679 thousands of USD, while we know that their true value is 3.500 thousands of USD! So the actual quarterly fixed costs exceed the obtained estimate by more than 100%! Well, it can definitely be described as significant understatement!

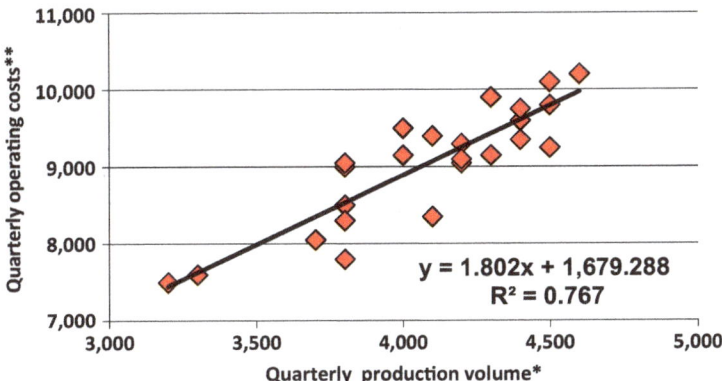

Fig. 6.9 The statistical relationship between company-level quarterly total operating costs and company-level quarterly production volume (on the basis of data from the last two columns of Table 6.23). *For the whole company, in physical units. **For the whole company, in thousands of USD. Source: authors

If your task is to evaluate the company's operating risk on the grounds of the estimated share of fixed operating costs in total operating costs, then you would probably divide the estimated quarterly fixed costs by the average total quarterly costs. The arithmetic average of total quarterly costs, based on the data from the next to last column of Table 6.23, is USD 9.031 thousand. Therefore, the estimated average share of fixed costs in total costs is 18,6% (1.679/9.031), while the actual share is 38,8% (3.500/9.031). Quite a difference, resulting in probable understatement of company's operating risks.

Actual parameters of cost functions at individual factory level were constant throughout the whole 24-quarter period while the company-level costs constituted simple sums of factory-level costs. We can conclude, therefore, that there were no structural changes of the parameters of the relationships between operating costs and production volume. We also remember that the actual costs were generated in a deterministic way, as linear functions of production volumes in individual factories, without any residuals. But despite all of this, some factors have brought about the heavy underestimation of company-level quarterly fixed costs in our regression. There was one factor behind this: the inter-factory changes of production structure from quarter to quarter. Table 6.24 presents the shares of the individual manufacturing factories in total production of the company in individual quarters.

As can be seen in Table 6.24, the inter-factory breakdown of total production volume was variable in the course of the investigated 24 quarters. Each of three factories showed the highest share in some quarters. If we keep in mind how diverse these factories are in terms of their cost structures, then we are not surprised that such dynamically variable inter-factory production structure could heavily bias the parameters of the our cost regression, estimated on aggregated data. As we can see

Table 6.24 Shares of individual manufacturing factories in total production of the whole company

Number of quarter	Factory A (%)	Factory B (%)	Factory C (%)
1	26,32	31,58	42,11
2	29,27	24,39	46,34
3	48,78	17,07	34,15
4	37,21	25,58	37,21
5	34,21	34,21	31,58
6	36,84	36,84	26,32
7	26,32	28,95	44,74
8	24,44	33,33	42,22
9	18,60	46,51	34,88
10	25,00	42,50	32,50
11	29,55	34,09	36,36
12	37,50	37,50	25,00
13	22,50	32,50	45,00
14	36,36	25,00	38,64
15	50,00	23,68	26,32
16	37,78	31,11	31,11
17	33,33	35,71	30,95
18	39,39	33,33	27,27
19	33,33	23,81	42,86
20	27,27	34,09	38,64
21	19,57	50,00	30,43
22	26,67	40,00	33,33
23	30,95	40,48	28,57
24	40,54	32,43	27,03

Source: Authors

in the last column of Table 6.23, there were several quarters with total company-level production of 3.800 units. However, despite having the same total volume of production, these quarters differed significantly in terms of locations of production. It is obvious that production of 3.800 units in the first quarter (with over 42% of production volume in Factory C, which is characterized by the lowest fixed costs and the highest unit variable costs) results in different cost structure than production of 3.800 units in the 15th quarter (with 50% of production volume in Factory A, which has the highest share of fixed costs).

As we can see, even without any structural changes of parameters of the modeled relationships, such structural changes occurring *within* the variables in the model may strongly distort the regression parameters, if the regression is estimated on aggregated data. Moreover, in many cases these distortions are not signaled by any tests applied during verification of the model, as was the case with structural breaks occurring in the relationships between variables. Table 6.25 presents the underlying

Table 6.25 Observations, fitted values, and residuals of the regression of company-level total operating costs against company-level production volume

Number of quarter	TOC[a] (dependent variable)	PV[b] (explanatory variable)	Fitted values of dependent variable	Residuals
1	9.000	3.800	8.528,10	471,90
2	9.400	4.100	9.068,80	331,20
3	8.350	4.100	9.068,80	−718,80
4	9.150	4.300	9.429,26	−279,26
5	8.500	3.800	8.528,10	−28,10
6	8.300	3.800	8.528,10	−228,10
7	9.050	3.800	8.528,10	521,90
8	10.100	4.500	9.789,73	310,27
9	9.900	4.300	9.429,26	470,74
10	9.150	4.000	8.888,57	261,43
11	9.600	4.400	9.609,49	−9,49
12	7.500	3.200	7.446,71	53,29
13	9.500	4.000	8.888,57	611,43
14	9.350	4.400	9.609,49	−259,49
15	7.800	3.800	8.528,10	−728,10
16	9.250	4.500	9.789,73	−539,73
17	9.050	4.200	9.249,03	−199,03
18	7.600	3.300	7.626,94	−26,94
19	9.300	4.200	9.249,03	50,97
20	9.750	4.400	9.609,49	140,51
21	10.200	4.600	9.969,96	230,04
22	9.800	4.500	9.789,73	10,27
23	9.100	4.200	9.249,03	−149,03
24	8.050	3.700	8.347,87	−297,87

Source: Authors
[a]TOC = total company-level operating costs in a given quarter in thousands of USD
[b]PV = company-level production volume in a given quarter in physical units

data as well as fitted values and residuals from our regression of company-level total operating costs against company-level production volume. Table 6.26 in turn shows the results of autocorrelation, heteroscedasticity ,and Ramsey's *RESET* tests conducted for this regression. As can be seen, none of the applied tests suggested any incorrectness of the estimated regression. Therefore, such statistical tools cannot warn us against the incredibility of the regression parameters caused by structural changes occurring within the individual variables included in the model. Only analyst's combined economic knowledge, experience, and common sense may be helpful here, emphasizing the importance of nonstatistical knowledge in statistical analyses.

Table 6.26 Tests for autocorrelation, heteroscedasticity, and specification tests conducted for the regression of company-level total operating costs against company-level production volume

F-test for autocorrelation of residuals (of up to fourth order)	
R-squared of regression of residuals against lagged (up to fourth order) residuals	0,291
Empirical *F* statistic (for $n = 20$ and $k = 4$)	1,54
Critical value of *F* statistic	3,06
Conclusion:	Residuals are not autocorrelated
ARCH-LM test for heteroscedasticity of residuals (of up to fourth order)	
R-squared of regression of squared residuals against lagged (up to fourth order) squared residuals	0,066
Empirical *F* statistic (for $n = 20$ and $k = 4$)	0,26
Critical value of *F* statistic	3,06
Conclusion:	Residuals are homoscedastic
Ramsey's *RESET* test for regression specification	
R-squared of tested regression (the same as in Fig. 6.9)	0,767
R-squared of auxiliary regression (with two auxiliary variables[a])	0,767
Empirical *F* statistic	0,00
Critical value of *F* statistic	3,49
Conclusion:	Regression is correctly specified

Source: Authors
[a]Two auxiliary variables were added to the tested regression: squared fitted values of dependent variable and cubic fitted values of dependent variable

Reference

Granger CWJ, Newbold P (1986) Forecasting economic time series. Academic Press, Orlando

Chapter 7
Regression Analysis of Discrete Dependent Variables

7.1 The Nature and Examples of Discrete Dependent Variables

Many social and economic variables are of a qualitative nature. They are examples of discrete variables. These variables relate mainly to events or processes that result in a small number of possibilities, where many times the key issue is whether the particular event happens or not.

Among examples of discrete variables often investigated in marketing, economic and finance research, are:

- Occurrence of a recession.
- Company bankruptcy.
- Purchase by a given consumer the product or service in a given time.
- Maintenance of the consumer's loyalty toward a given brand.
- Timely payment of debt and interest by the borrower.
- Customer's decision to live in an owned (vs. rented) apartment.
- Customer's habit to travel to a workplace by public bus (instead of a private car).
- Customer's habit to spend a holiday abroad (vs. in a home country).
- Customer's habit to travel abroad on his or her own (instead of buying holidays from a travel agent).

The existence of such variables means that often in economic analyses, the key question is "whether?" or "if?", instead of "by how much?", and the considered answers are of zero-one (binary) character: "yes" or "not." Therefore, if in statistical research the explained (or predicted) variable is of a binary nature, then the individual observations of this variable have the values of only 1 (something has happened or is going to happen) or 0 (something has not happened or is not expected to happen).

© Springer International Publishing AG 2018
J. Welc, P.J.R. Esquerdo, *Applied Regression Analysis for Business*,
https://doi.org/10.1007/978-3-319-71156-0_7

The dependent variables of this kind can be modeled by several tools, including (among others):

- Discriminant analysis.
- Logit function.

7.2 The Discriminant Analysis

7.2.1 Nature and Estimation of Discriminant Models

The idea of discriminant analysis boils down to the construction of an econometric model capable of "discriminating" (dividing) all the observations under investigation into one or more nonoverlapping groups (Hardle and Simar 2007). In the case of two groups, individual observations are assigned either to the group for which the event occurred (e.g., the customer agreed to a new contract during the phone call from the seller) or to the group for which the event did not occur (e.g., the customer, despite being presented the new promotional offer with details, did not agree to sign the new contract).

Similarly as in the case of the traditional econometric models, the discriminant models include:

- A dependent variable (having the values of only 1 or 0).
- A set of explanatory variables (which can have both continuous or binary character), with statistically significant relationships with the dependent variable.

There are many alternative methods of estimating the parameters of the discriminant functions; however, for illustrative purposes we will present only one of the simplest.

The construction of the discriminating models boils down to estimating the parameters of the following linear equation:

$$Y = \alpha_1 X_1 + \alpha_2 X_2 + \ldots + \alpha_n X_n$$

where:

$$Y = \begin{cases} 1 : \text{if the analyzed event has happened} \\ 0 : \text{if the analyzed event has not happened} \end{cases}$$

$X_1, X_2, \ldots, X_n$—the set of n explanatory variables, showing statistically significant relationships with the dependent variable Y,

$\alpha_1, \alpha_2, \ldots, \alpha_n$—the parameters of the discriminating function (which are to be estimated on the basis of the empirical data relating to the dependent variable as well as the explanatory variables).

The vector of n parameters $\alpha_1, \alpha_2, \ldots, \alpha_n$ can be obtained by solving the following equations:

$$Ca = g$$

where:

$$C = C_1 + C_2$$

C_1, C_2—variance and covariance matrices in both groups formed on the basis of the values of the dependent variable,

$$g = g_1 - g_2$$

g_1, g_2—vector of arithmetic averages of the individual explanatory variables in both groups formed on the basis of the values of the dependent variable,

$$\alpha = gC^{-1}.$$

Finally, in order to enable the classification of any given observation into one of two alternative groups, the threshold value of the discriminating function f_0 must be determined. The simplest way of doing this is by using the following formula:

$$f_0 = \frac{\bar{f}_1 + \bar{f}_2}{2}$$

where $\bar{f}_1$ and $\bar{f}_2$ denote the arithmetic averages of theoretical (fitted) values of the discriminating function computed for the observations belonging to both groups.

As in the case of multiple regression, the important issue is the selection of the explanatory variables. One of the simplest methods for choosing the explanatory variables in the case of discriminant analysis is based on the comparison of the average values of a given candidate explanatory variable in two classification groups. If a given variable is significant in discriminating, then its average values in both groups should differ significantly. Formally, such an analysis should be done by means of the test of equality of two samples arithmetic averages or medians. However, for illustrative purposes, we will use the simplified approach, based on comparing the percent differences between the two averages.

7.2.2   Example of an Application of a Discriminant Function

A telecommunications company prepares a new promotional campaign aimed at boosting sales of new services (e.g., TV and Internet services). The campaign is going to be based on direct phone calls made to targeted consumers, which already belong to the company's consumer base. The purpose of these phone calls is to present the new offer to the potential buyers and to sell as many new services as possible during these first phone calls.

Theoretically, the company can maximize the sales effects of this marketing campaign by calling all its current customers. However, this is very expensive and requires costly phone calls to perhaps millions of would-be-buyers, which would bring huge telecommunication costs as well as huge amounts paid to an army of marketers. Also, this would be very time consuming, because it's not possible to call millions of customers in a short time. Therefore, in business practice, the more economic (and sometimes the only feasible) strategy is to identify those customers who bear the highest probability of success, which in this case means a high probability of being convinced to buy a new service during the first phone call from the company.

If the company conducted similar marketing campaigns in the past, it should have some database about the achieved results (e.g., the number and data of customers who purchased the services). Such a database can be used for trying to establish relationships between the customers' decisions in the past (i.e., whether they purchased the new services during the phone call or whether they rejected the new offer) and the customers' personal features (such as age, sex, income, etc.).

Suppose that the company has already run such a marketing campaign in the past (at that time by calling to some customers selected at random) and as a result it has a database related to the 15 customers which purchased the service, and 15 which rejected the offer. For building a discriminant function, the number of observations in both groups can differ. However, for illustrative purposes, we assume that they are equal.

The company analysts hypothesize that some of the consumers' personal features are correlated with these consumers' decisions (specifically, the consumers' willingness to buy the new service during the first phone call). The analysts proposed the following set of candidate explanatory factors:

- Consumer's age (in years).
- Consumer's sex (denoted as 1 if female and 0 if male).
- Consumer's average monthly bill for the telecommunication services purchased from the company (in USD).
- Consumer's place of living (denoted as 1 if living in the town/city and 0 if living in the rural area).
- Consumer's marital status (denoted as 1 if single and 0 if married).
- Consumer's average monthly income (in USD).
- Consumer's parental status (denoted as 1 if living with children and 0 if not living with any children).
- Consumer's higher education status (denoted as 1 if higher and 0 otherwise).
- Consumer's primary education (denoted as 1 if consumer ceased education at the primary level and 0 otherwise).
- Consumer's religion (denoted as 1 if atheist and 0 otherwise).

The proposed set of the candidate explanatory variables entails some problems related to data availability. Even though variables such as religion, level of education, or average incomes can have a clear impact on the consumers' decisions, the companies rarely can gather data on them legally. Suppose that this company has only easily obtainable data related to the first four variables (i.e., age, sex, average monthly bill, and place of living). The data from the previous marketing campaign are presented in Tables 7.1 and 7.2.

Table 7.1 Data for 15 consumers who purchased the new service in the course of the phone call during the previous marketing campaign

Number of consumers	Consumer's age (in years)	Consumer's sex (1—female, 0—male)	Consumer's average monthly bill in USD	Consumer's place of living (1—city/town, 0—rural area)
1	20	1	95	0
2	18	0	120	0
3	30	1	150	1
4	33	1	110	0
5	28	0	55	0
6	32	1	72	0
7	19	1	96	1
8	24	0	105	0
9	19	1	83	0
10	25	1	160	1
11	32	1	144	0
12	34	0	91	0
13	42	0	115	0
14	40	1	68	0
15	45	1	71	0

Source: Authors

Table 7.2 Data for 15 consumers who did not purchase the new service in the course of the phone call during the previous marketing campaign

Number of consumers	Consumer's age (in years)	Consumer's sex (1—female, 0—male)	Consumer's average monthly bill in USD	Consumer's place of living (1—city/town, 0—rural area)
16	26	1	145	1
17	24	0	60	1
18	28	0	100	0
19	35	1	110	1
20	34	1	112	1
21	40	0	140	0
22	35	0	115	1
23	45	0	80	1
24	50	1	73	1
25	62	0	120	0
26	43	0	80	1
27	45	0	95	1
28	39	0	100	0
29	51	1	70	1
30	55	0	157	0

Source: Authors

Table 7.3 Selection of explanatory variables for a discriminant analysis

	Consumer's age (in years)	Consumer's sex (1—female, 0—male)	Consumer's average monthly bill in USD	Consumer's place of living (1—city/town, 0—rural area)
Arithmetic average for consumers who purchased the service during the phone call ($Y = 1$)	29,40	0,67	102,33	0,20
Arithmetic average for consumers who did not purchase the service during the phone call ($Y = 0$)	40,80	0,33	103,80	0,67
Percent difference ($Y = 1/Y = 0$)	−27,9%	100,0%	−1,4%	−70,0%
Percent difference ($Y = 0/Y = 1$)	38,8%	−50,0%	1,4%	233,3%

Source: Authors

The first step is the selection of the explanatory variables. We consider the set of four candidate explanatory variables. However, some of them can be statistically insignificant, bringing unwelcome noise into our model. Therefore, we include in the model only those variables which have significant statistical relationships with the dependent variable.

The selection of the explanatory variables can be done by comparing the average values of all the candidate explanatory variables in both classification groups. Table 7.3 presents such a comparison.

As the above data show, the arithmetic averages differ significantly between the two groups in three out of the four candidate variables (i.e., consumer's age, consumer's sex, and consumer's place of living).

The directions of these differences suggest that:

- The consumers who purchased the service are characterized by significantly younger average age (29,40 years vs. 40,80 years).
- The consumers who purchased the service are characterized by significantly larger average value of "Sex" variable (0,67 vs. 0,33), which means that women have higher propensity to purchase, on average.
- The consumers who purchased the service are characterized by significantly lower average value of "Residence" (place of living) variable (0,20 vs. 0,67), which means that city/town inhabitants are less willing to purchase during the phone call, on average.

However, the remaining candidate variable (i.e., consumer's average monthly bill) seems to have very similar average values in both groups (the differences between the averages are lower that ±1,5%). The average bill among those consumers, who purchased the new service during the phone call in the previous marketing campaign, amounts to USD 102,33 per month, as compared to USD 103,80 per month in the group of consumers who rejected the offer. It suggests, therefore, that this variable does not help much in discriminating the consumers with high propensity to purchase from those with low propensity. The variable can be removed from further analysis.

Thus, our task is to estimate the parameters of the following linear equation:

$$Y = \alpha_1 Age + \alpha_2 Sex + \alpha_3 Residence$$

where:

$$Y = \begin{cases} 1 : \text{if the given consumer purchased the new service during the phone call} \\ 0 : \text{if the given consumer did not purchase the new service during the phone call} \end{cases}$$

Age—consumer's age (in years),
Sex—consumer's sex (1—female, 0—male),
Residence—consumer's place of living (1—city/town, 0—rural area),
$\alpha_1, \alpha_2, \alpha_3$—parameters of the discriminating function (which are to be estimated).

The first step, according to the estimation procedure described earlier, is to compute the vectors g_1, g_2 and vector g (which is the difference between g_1 and g_2) as follows:

$$g_1 = [29, 40 \quad 0, 67 \quad 0, 20],$$

$$g_2 = [40, 80 \quad 0, 33 \quad 0, 67],$$

$$g = [-11, 40 \quad 0, 33 \quad -0, 47],$$

The next step is to compute the variance and covariance matrices C_1, C_2 and the matrix C (which is the sum of C_1 and C_2) as follows:

$$C_1 = \begin{bmatrix} 69, 84 & 0, 07 & -0, 95 \\ 0, 07 & 0, 22 & 0, 07 \\ -0, 95 & 0, 07 & 0, 16 \end{bmatrix}$$

$$C_2 = \begin{bmatrix} 110, 83 & -0, 53 & -1, 33 \\ -0, 53 & 0, 22 & 0, 11 \\ -1, 33 & 0, 11 & 0, 22 \end{bmatrix}$$

$$C = \begin{bmatrix} 180, 67 & -0, 47 & -2, 28 \\ -0, 47 & 0, 44 & 0, 18 \\ -2, 28 & 0, 18 & 0, 38 \end{bmatrix}$$

Then the estimation procedure requires computation of the inverse of the matrix C:

$$C^{-1} = \begin{bmatrix} 0,006 & -0,010 & 0,041 \\ -0,010 & 2,781 & -1,352 \\ 0,041 & -1,352 & 3,487 \end{bmatrix}$$

Finally, we compute the vector of the parameters, by using the following formula:

$$\alpha = gC^{-1}.$$

$$\alpha = [-11,40 \quad 0,33 \quad -0,47] \begin{bmatrix} 0,006 & -0,010 & 0,041 \\ -0,010 & 2,781 & -1,352 \\ 0,041 & -1,352 & 3,487 \end{bmatrix}$$

$$\alpha = [-0,091 \quad 1,671 \quad -2,540].$$

Putting the estimated parameters into our discriminating function gives us the following model:

$$Y = -0,091 \, Age + 1,671 \, Sex - 2,540 \, Residence.$$

The obtained signs of the model parameters inform us that:

- The relationship between the consumers' age and their propensity to purchase the service during the phone call is negative (meaning that the older the given consumer, the lower the probability of making a purchase during the phone call from the company).
- The relationship between the consumers' sex and their propensity to purchase the service during the phone call is positive (meaning that women are more willing to buy during the phone call).
- The relationship between the consumers' residence (place of living) and their propensity to purchase the service during the phone call is negative (meaning that city/town inhabitants are less willing to buy during the phone call).

The obtained discriminating function enables computation of the theoretical (fitted) values of Y for all the observations. This is shown in Tables 7.4 and 7.5.

Finally, in order to enable the classification of the consumers into one of two alternative groups, the threshold value of the discriminating function f_0 must be determined. According to the following formula:

$$f_0 = \frac{\bar{f}_1 + \bar{f}_2}{2},$$

where $\bar{f}_1$ and $\bar{f}_2$ denote the arithmetic averages of theoretical (fitted) values of the discriminating function, calculated and presented in the last rows of the above tables:

Table 7.4 Fitted values of the estimated discriminating function, computed for 15 consumers who purchased the new service

Number of consumers	Age[a] (−0,091)	Sex[a] (1,671)	Residence[a] (−2,540)	Theoretical (fitted) value f_1
1	20	1	0	−0,15
2	18	0	0	−1,64
3	30	1	1	−3,59
4	33	1	0	−1,33
5	28	0	0	−2,54
6	32	1	0	−1,24
7	19	1	1	−2,60
8	24	0	0	−2,18
9	19	1	0	−0,06
10	25	1	1	−3,14
11	32	1	0	−1,24
12	34	0	0	−3,09
13	42	0	0	−3,82
14	40	1	0	−1,96
15	45	1	0	−2,42
Arithmetic average $\bar{f}_1$				−2,06

Source: Authors
[a]Values of the parameters of the discriminating function in parentheses

Table 7.5 Fitted values of the estimated discriminating function, computed for 15 consumers who did not purchase the new service

Number of consumers	Age[a] (−0,091)	Sex[a] (1,671)	Residence[a] (−2,540)	Theoretical (fitted) value f_2
16	26	1	1	−3,23
17	24	0	1	−4,72
18	28	0	0	−2,54
19	35	1	1	−4,05
20	34	1	1	−3,96
21	40	0	0	−3,63
22	35	0	1	−5,72
23	45	0	1	−6,63
24	50	1	1	−5,41
25	62	0	0	−5,63
26	43	0	1	−6,45
27	45	0	1	−6,63
28	39	0	0	−3,54
29	51	1	1	−5,50
30	55	0	0	−5,00
Arithmetic average $\bar{f}_2$				−4,84

Source: Authors
[a]Values of the parameters of the discriminating function in parentheses

$$f_0 = \frac{(-2,06) + (-4,84)}{2} = -3,45.$$

According to our computations:

- Any customer for whom the calculated value of the discriminating function exceeds −3,45 should be classified as bearing relatively high probability of purchasing the offered new service during the phone call from the company's sales officer.
- Any customer for whom the calculated value of the discriminating function is below −3,45 should be classified as bearing relatively low probability of purchasing the offered new service during the phone call from the company's sales officer.

Table 7.6 repeats the values of the discriminating function, computed for the individual consumers in our sample, with additional information about the correctness of the classification.

As we can see, our discriminat model correctly classified 26 out of 30 observations (about 86,7%). It incorrectly classified two customers who purchased the service and another two, who did not purchase the service. The approximate classification error rate is about 13,3%.

Table 7.6 The analysis of the classification correctness based on the estimated discriminant function

Consumers who purchased the new service ($Y = 1$)			Consumers who did not purchase the new service ($Y = 0$)		
Number of consumers	Theoretical (fitted) value f_1	Is classification correct?	Number of consumers	Theoretical (fitted) value f_2	Is classification correct?
1	−0,15	Yes	16	−3,23	**No**
2	−1,64	Yes	17	−4,72	Yes
3	−3,59	**No**	18	−2,54	**No**
4	−1,33	Yes	19	−4,05	Yes
5	−2,54	Yes	20	−3,96	Yes
6	−1,24	Yes	21	−3,63	Yes
7	−2,60	Yes	22	−5,72	Yes
8	−2,18	Yes	23	−6,63	Yes
9	−0,06	Yes	24	−5,41	Yes
10	−3,14	Yes	25	−5,63	Yes
11	−1,24	Yes	26	−6,45	Yes
12	−3,09	Yes	27	−6,63	Yes
13	−3,82	**No**	28	−3,54	Yes
14	−1,96	Yes	29	−5,50	Yes
15	−2,42	Yes	30	−5,00	Yes

Source: Authors

However, the classification done above was based on the very strict critical value of the discriminating function. This means that two hypothetical customers, having discriminating function values of, e.g., −3,40 and −3,50, respectively, would be classified differently (although they seem to be very similar). Here, customer number 3 has the function value of −3,59 and is incorrectly classified as having high propensity to buy while the customer number 28 has the function value of −3,54 and is correctly classified as having low propensity to buy.

Therefore, it is usually justified to create some fuzzy line between the two classification groups. For example, the analyst could assume the arbitrary "gray area" bounded by the critical value ±10%. Here, it would give the "gray area" between −3,80 and −3,11. If the risk or the known negative consequence of incorrect classification is high, then only the observations which are outside the "gray area" should be classified (and those within the "gray area" should be considered indeterminate). The higher the risk or negative consequence of incorrect classification, the wider the "gray area" should be.

The constructed discriminat model can be used for forecasting the behavior of other consumers, not only those who constituted the sample used in the model estimation. However, any user of such a model should always keep in mind that its out-of-sample predictive accuracy (i.e., the ability to forecast the behavior of objects from outside the estimation sample) may be worse than its in-sample goodness of fit.

7.3 The Logit Function

7.3.1 Nature and Estimation of Logit Models

The discriminant function presented in the previous section classified the customers into one of two groups based on the obtained value of the discriminant function for each individual customer. The discriminant function generates values of the dependent variable which are not directly interpretable. For example, the value obtained for the 15th customer in Table 7.6 (−2,42) is itself meaningless and non-interpretable. To obtain any finding, we must compare that value with the threshold computed for the classification.

An alternative tool to the discriminant analysis is the logit model, in which the obtained value of the dependent variable is the probability of the event (e.g., the customer decision to purchase a product), bounded in a range between 0% and 100% (Cox 1959; Freedman 2009). The advantage of the logit model over the discriminant analysis lies in the more understandable and clear suggestions about the classification of a given object. For example, if the estimated probability of a recession is, say, 70%, we may immediately conclude (without any reference to the formally computed thresholds) that the recession is more likely than not.

As in the case of the discriminant function, the dependent variable in a logit model may (but does not have to) be a binary variable. However, while in the discriminant function, the dependent variable entered into the model in its binary

Table 7.7 The procedure of logit model estimation for a binary dependent variable

Step 1	Transform a binary dependent variable Y of the following form: $Y = \begin{cases} 1 : \text{if the analyzed event has happened} \\ 0 : \text{if the analyzed event has not happened} \end{cases}$ into the probability variable P of the following form: $P = \begin{cases} 0,999 : & \text{if the analyzed event happened} \\ 0,001 : & \text{if the analyzed event has not happened} \end{cases}$
Step 2	Transform each observation of a probability variable P into a logit variable L with the following formula: $L_i = \log\frac{p_i}{1-p_i}$, where: L_i—value of logit for i-the observation, p_i—probability of an event in the case of i-the observation.
Step 3	Estimate the regression of the logit variable L against the set of its explanatory variables and compute its fitted values.
Step 4	Reverse the fitted values of individual observations of the logit variable L into the probabilities with the following formula: $p_i = \frac{10^{L_i}}{1+10^{L_i}}$

Source: Authors

form, for the estimation of the logit function, the binary variable is transformed into so-called logits. The procedure of estimation and application of the logit model for a binary variable is presented in Table 7.7.

Note that in Step 1, the two variants of the binary variable are transformed into the probabilities which do not equal exactly 0% or 100%. Instead, the observations in which the analyzed event has happened are transformed into the very high probability of 99,9%, while the observations for the cases in which the event has not happened are transformed into the very low probability of 0,01%. This is a simplification stemming from the fact that the logit transformation in Step 2 is mathematically unfeasible for probabilities equaling exactly 0% and 100%. The applied simplification enables computation of logits for almost 0% and almost 100% probabilities without significant distortions of the analytical findings.

7.3.2 Example of an Application of a Logit Model

We illustrate the application of the logit model with the same example as before, where we used discriminant analysis. Tables 7.1 and 7.2 above presented the data for 15 consumers who purchased the new service and 15 consumers who did not purchase it, respectively, in the course of the phone call during the previous marketing campaign.

First, we define our binary dependent variable Y as follows:

$$Y = \begin{cases} 1 : \text{if the given consumer purchased the new service during the phone call} \\ 0 : \text{if the given consumer did not purchase the new service during the phone call} \end{cases}$$

and then we transform it into the probability variable P of the following form:

$$P = \begin{cases} 0,999 : \text{if the given consumer purchased the new service during the phone call} \\ 0,001 : \text{if the given consumer did not purchase the new service during the phone call} \end{cases}$$

Now we proceed to Step 2 and transform each observation of the probability variable P into a logit variable L. The logit transformations of both values of the probability variable P look as follows:

- For probability of 99,9% (the value of the probability variable P of 0,999):

$$L_{p=0,999} = \log\frac{0,999}{1 - 0,999} = 3,0.$$

- For probability of 0,01% (the value of the probability variable P of 0,001):

$$L_{p=0,001} = \log\frac{0,001}{1 - 0,001} = -3,0.$$

Table 7.8 presents the values of our dependent variable, which is the logit variable L, together with the observations of its three explanatory variables, which are the same as in the discriminant function estimated in the previous section. The table also presents the fitted values of the dependent variable, obtained after regressing the logit variable against its three drivers.

Table 7.9, in turn, presents the results of regressing the logit variable against its three explanatory variables. As we can see, the estimated model is statistically significant, given its F statistic. Also, all its three explanatory variables are statistically significant. Although the R-squared statistic of the estimated logit regression is not very high, it seems to suggest that the obtained model has some explanatory power.

Now we may proceed to computing the fitted probabilities of purchasing the new service by the individual consumers within the sample. To this end we need to use the data from the last column of Table 7.8 and the formula presented in Step 4 in Table 7.7. The obtained probabilities as well as the classifications of the individual consumers are shown in Table 7.10. For classifying the consumers, we assumed arbitrarily that the threshold of a probability equals 50%.

The example of a computation of a probability of purchase of a service, on the basis of the fitted value of a logit function, for the tenth consumer (for whom the fitted logit $L_{10} = 0,394$), looks as follows:

$$p_{10} = \frac{10^{L_{10}}}{1 + 10^{L_{10}}} = \frac{10^{0,394}}{1 + 10^{0,394}} = 71,2\%.$$

Similarly as in the case of the discriminant function illustrated in the preceding section, the estimated logit model correctly classified 26 out of 30 observations (about 86,7%). It incorrectly classified two customers who purchased the service

Table 7.8 Actual and fitted values of the logit function

Number of consumers		Actual value of a logit variable	Age[a] (−0,114)	Sex[a] (2,098)	Residence[a] (−3,190)	Fitted value of a logit variable
Consumers who purchased the new service	1	3,0	20	1	0	4,154
	2	3,0	18	0	0	2,284
	3	3,0	30	1	1	−0,177
	4	3,0	33	1	0	2,671
	5	3,0	28	0	0	1,143
	6	3,0	32	1	0	2,785
	7	3,0	19	1	1	1,078
	8	3,0	24	0	0	1,599
	9	3,0	19	1	0	4,268
	10	3,0	25	1	1	0,394
	11	3,0	32	1	0	2,785
	12	3,0	34	0	0	0,459
	13	3,0	42	0	0	−0,454
	14	3,0	40	1	0	1,872
	15	3,0	45	1	0	1,302
Consumers who did not purchase the new service	16	−3,0	26	1	1	0,280
	17	−3,0	24	0	1	−1,590
	18	−3,0	28	0	0	1,143
	19	−3,0	35	1	1	−0,747
	20	−3,0	34	1	1	−0,633
	21	−3,0	40	0	0	−0,226
	22	−3,0	35	0	1	−2,845
	23	−3,0	45	0	1	−3,986
	24	−3,0	50	1	1	−2,458
	25	−3,0	62	0	0	−2,736
	26	−3,0	43	0	1	−3,758
	27	−3,0	45	0	1	−3,986
	28	−3,0	39	0	0	−0,112
	29	−3,0	51	1	1	−2,572
	30	−3,0	55	0	0	−1,937

Source: Authors
[a]Values of the parameters of the logit model in parentheses (the full estimation results are presented in Table 7.9)

and another two, who did not purchase the service. The consumers incorrectly classified by the logit model are the same as in the case of the discriminant function (which does not have to be the case).

Table 7.9 The parameters of a logit model, estimated on the basis of data from Table 7.8

Variable	Coefficient	t-Statistic
Intercept	4,337	3,030
Age	−0,114	−3,234
Sex	2,098	2,630
Residence	−3,190	−4,053

Critical value of t-Statistic (at 5% significance level): 2,056
R-squared: 0,581
F statistic: 12,04
Critical value of F statistic (at 5% significance level): 2,98

Source: Authors

Table 7.10 The analysis of the classification correctness based on the estimated logit model

Consumers who purchased the new service				Consumers who did not purchase the new service			
Num. of cons.	Fitted value of a logit variable	Estimated probability (%)	Is classification correct?[a]	Num. of cons.	Fitted value of a logit variable	Estimated probability (%)	Is classification correct?[a]
1	4,154	100,0	Yes	16	0,280	65,6	No
2	2,284	99,5	Yes	17	−1,590	2,5	Yes
3	−0,177	39,9	No	18	1,143	93,3	No
4	2,671	99,8	Yes	19	−0,747	15,2	Yes
5	1,143	93,3	Yes	20	−0,633	18,9	Yes
6	2,785	99,8	Yes	21	−0,226	37,3	Yes
7	1,078	92,3	Yes	22	−2,845	0,1	Yes
8	1,599	97,5	Yes	23	−3,986	0,0	Yes
9	4,268	100,0	Yes	24	−2,458	0,3	Yes
10	0,394	71,2	Yes	25	−2,736	0,2	Yes
11	2,785	99,8	Yes	26	−3,758	0,0	Yes
12	0,459	74,2	Yes	27	−3,986	0,0	Yes
13	−0,454	26,0	No	28	−0,112	43,6	Yes
14	1,872	98,7	Yes	29	−2,572	0,3	Yes
15	1,302	95,2	Yes	30	−1,937	1,1	Yes

Source: Authors
[a]An arbitrary 50% threshold of probability has been assumed for the classification

References

Cox DR (1959) The regression analysis of binary sequences (with discussion). J R Stat Soc B 20:215–242
Freedman DA (2009) Statistical models: theory and practice. Cambridge University Press, Cambridge
Hardle W, Simar L (2007) Applied multivariate statistical analysis. Springer, Berlin

Chapter 8
Real-Life Case Study: The Quarterly Sales Revenues of Nokia Corporation

8.1 Introduction

In this chapter of the book, we offer an example of a regression model which may be useful in predicting company's quarterly sales revenues. The presented example constitutes a real-life case study, based on time-series data reported by Nokia Corporation in its quarterly financial reports for the individual quarters of 2002–2011 period.

We will illustrate individual phases of model building and its application to forecasting. First, we propose a preliminary set of explanatory variables and check for the presence of any potential univariate and multivariate outliers in the dataset. Then we apply the procedure of "general to specific" modeling to select the final set of explanatory variables and test for its statistical correctness. Finally, we close the chapter with the evaluation of the predictive capabilities of the obtained model.

8.2 Preliminary Specification of the Model

We are interested in building and testing a simple autoregressive model which could be useful in predicting Nokia's sales revenues with one-quarter-ahead forecast horizon. To this end, we extracted sales revenue data from Nokia's quarterly financial statements, in a 10-year period beginning in the first quarter of 2002 and ending in the fourth quarter of 2011. These data are presented in Table 8.1.

The model is intended to be used in forecasting Nokia's future quarterly sales revenues after 2011. However, before the model is applied in forecasting, its predictive capabilities need to be checked with the sample of historical data. Therefore, we split the whole sample depicted in Table 8.1 into two subsamples:

© Springer International Publishing AG 2018
J. Welc, P.J.R. Esquerdo, *Applied Regression Analysis for Business*,
https://doi.org/10.1007/978-3-319-71156-0_8

Table 8.1 Quarterly sales revenues of Nokia Corporation in 2002–2011

Period	Nokia's quarterly sales revenues in EUR million	Period	Nokia's quarterly sales revenues in EUR million
I q. 2002	7.014	I q. 2007	9.856
II q. 2002	6.935	II q. 2007	12.587
III q. 2002	7.224	III q. 2007	12.898
IV q. 2002	8.843	IV q. 2007	15.717
I q. 2003	6.773	I q. 2008	12.660
II q. 2003	7.019	II q. 2008	13.151
III q. 2003	6.874	III q. 2008	12.237
IV q. 2003	8.789	IV q. 2008	12.662
I q. 2004	6.625	I q. 2009	9.274
II q. 2004	6.640	II q. 2009	9.912
III q. 2004	6.939	III q. 2009	9.810
IV q. 2004	9.063	IV q. 2009	11.988
I q. 2005	7.396	I q. 2010	9.522
II q. 2005	8.059	II q. 2010	10.003
III q. 2005	8.403	III q. 2010	10.270
IV q. 2005	10.333	IV q. 2010	12.651
I q. 2006	9.507	I q. 2011	10.399
II q. 2006	9.813	II q. 2011	9.275
III q. 2006	10.100	III q. 2011	8.980
IV q. 2006	11.701	IV q. 2011	10.005

Source: Nokia's quarterly financial statements

- Estimation subsample—covering the period since the beginning of 2002 to the end of 2008 (28 observations).
- Forecast subsample—covering the period since the beginning of 2009 to the end of 2011 (12 observations).

The model will be estimated and verified on the basis of the first 28 observations. Then, if it satisfies all the crucial requirements for model correctness, including normally distributed residuals and lack of autocorrelation, it will be used in building forecasts of sales revenues for 12 individual quarters of 2009–2011 period. Finally, the average forecast errors will be computed and evaluated.

The dependent variable is Nokia's quarterly sales revenues (in millions of EUR), denoted as SR_t, where t denotes a particular quarter. In this autoregressive model, the candidate explanatory variables are the lags, up to the fourth quarter, of the dependent variable. However, the model is based on quarterly data, which calls for taking into consideration a possible seasonality of sales. Thus, we will add three seasonal dummy variables to the set of candidate explanatory variables. Before the model is estimated, two simplifying arbitrary assumptions are taken. First, the maximum lag of sales revenues is set at four quarters. Second, the seasonality (if present) is assumed to be of an additive nature.

The preliminary set of candidate explanatory variables are the following:

- SR_{t-1}—Nokia's quarterly sales revenues (in millions of EUR) lagged by one quarter.
- SR_{t-2}—Nokia's quarterly sales revenues (in millions of EUR) lagged by two quarters.
- SR_{t-3}—Nokia's quarterly sales revenues (in millions of EUR) lagged by three quarters.
- SR_{t-4}—Nokia's quarterly sales revenues (in millions of EUR) lagged by four quarters.
- *First*—dummy variable with value of unity in the first quarter and zero in remaining three quarters.
- *Second*—dummy variable with value of unity in the second quarter and zero in remaining three quarters.
- *Third*—dummy variable with value of unity in the third quarter and zero in remaining three quarters.

The initial set of data for the autoregressive model with additive seasonality is shown in Table 8.2. As always in the case of autoregressive models, some observations at the beginning of the sample are lost because of the lags of the dependent variable. Here, the maximum lag is set at four quarters, which results in the loss of four observations.

8.3 Detection of Potential Outliers in the Dataset

The next step in our procedure of econometric modeling is to check for the presence of any outliers in our dataset. In order to do this, we will apply two complementary tools:

- Apply the "two-sigma range" to the individual variables (except for seasonal dummies, which by their nature are free from outliers).
- Apply the "two-sigma range" to the residuals obtained from the general regression.

8.3.1 The "Two-Sigma Range" Applied to the Individual Variables

Our first step in searching for outliers is to apply the "two-sigma range" to the individual variables (except for dummies for seasonality). In order to reduce the risk of overlooking potential outliers, we set the upper and lower bounds based on their respective arithmetic means and two standard deviations. Such an approach results in narrower ranges (as compared to the "three-sigma rule") whereby more

Table 8.2 The statistical data for all the observations of the dependent variable and the candidate explanatory variables in the model for Nokia's quarterly sales revenues

Period	Dependent variable SR_t	Candidate explanatory variables SR_{t-1}	SR_{t-2}	SR_{t-3}	SR_{t-4}	First	Second	Third
I q. 2002	7.014	–	–	–	–	1	0	0
II q. 2002	6.935	7.014	–	–	–	0	1	0
III q. 2002	7.224	6.935	7.014	–	–	0	0	1
IV q. 2002	8.843	7.224	6.935	7.014	–	0	0	0
I q. 2003	6.773	8.843	7.224	6.935	7.014	1	0	0
II q. 2003	7.019	6.773	8.843	7.224	6.935	0	1	0
III q. 2003	6.874	7.019	6.773	8.843	7.224	0	0	1
IV q. 2003	8.789	6.874	7.019	6.773	8.843	0	0	0
I q. 2004	6.625	8.789	6.874	7.019	6.773	1	0	0
II q. 2004	6.640	6.625	8.789	6.874	7.019	0	1	0
III q. 2004	6.939	6.640	6.625	8.789	6.874	0	0	1
IV q. 2004	9.063	6.939	6.640	6.625	8.789	0	0	0
I q. 2005	7.396	9.063	6.939	6.640	6.625	1	0	0
II q. 2005	8.059	7.396	9.063	6.939	6.640	0	1	0
III q. 2005	8.403	8.059	7.396	9.063	6.939	0	0	1
IV q. 2005	10.333	8.403	8.059	7.396	9.063	0	0	0
I q. 2006	9.507	10.333	8.403	8.059	7.396	1	0	0
II q. 2006	9.813	9.507	10.333	8.403	8.059	0	1	0
III q. 2006	10.100	9.813	9.507	10.333	8.403	0	0	1
IV q. 2006	11.701	10.100	9.813	9.507	10.333	0	0	0
I q. 2007	9.856	11.701	10.100	9.813	9.507	1	0	0
II q. 2007	12.587	9.856	11.701	10.100	9.813	0	1	0
III q. 2007	12.898	12.587	9.856	11.701	10.100	0	0	1
IV q. 2007	15.717	12.898	12.587	9.856	11.701	0	0	0
I q. 2008	12.660	15.717	12.898	12.587	9.856	1	0	0
II q. 2008	13.151	12.660	15.717	12.898	12.587	0	1	0
III q. 2008	12.237	13.151	12.660	15.717	12.898	0	0	1
IV q. 2008	12.662	12.237	13.151	12.660	15.717	0	0	0

Source: Nokia's quarterly financial statements; authors

potential outliers may be captured, at the small risk of misidentifying some true observations as outliers. Measures for the "two-sigma range" (i.e., arithmetic means and standard deviations of individual variables) are computed on the time series covering the period between the first quarter of 2003 and the last quarter of 2008, excluding all observations from 2002, which are lost when lagging the dependent variable. Table 8.3 presents the results of the analysis.

Table 8.3 shows that in the case of all five investigated variables, the largest values in the sample exceed their respective upper bounds as determined by the "two-sigma range." However, for all five variables, the largest observations have identical value of 15.717. From Table 8.2, we can infer that this relates to Nokia's sales revenues reported for the fourth quarter of 2007, resulting from the spillover

Table 8.3 "Two-sigma range" applied to the Nokia's quarterly sales revenues as well as its lags

Measure	SR_t	SR_{t-1}	SR_{t-2}	SR_{t-3}	SR_{t-4}
Arithmetic mean (AM)	9.825,08	9.665,96	9.457,08	9.198,08	8.962,83
Std. deviation (SD)	2.604,46	2.539,45	2.524,90	2.447,13	2.370,02
AM + 2 × SD	15.034,00	14.744,85	14.506,89	14.092,34	13.702,88
AM − 2 × SD	4.616,17	4.587,06	4.407,28	4.303,83	4.222,79
Max in the sample	15.717	15.717	15.717	15.717	15.717
Min in the sample	6.625	6.625	6.625	6.625	6.625

Source: Nokia's quarterly financial statements; authors

caused by lagging the dependent variable. There are no other observations with values exceeding the upper bounds. In contrast, the lowest values of individual variables lie significantly above their respective lower bounds.

As we suggested in Chap. 5, the general principle in such autoregressions is to include in the set of candidate explanatory variables as many dummy variables as the number of observations "infected" by the outlying values. In our dataset, five consecutive observations (from the fourth quarter of 2007 to the fourth quarter of 2008) are "infected" by the outlying value of the dependent variable. Therefore, we should consider adding five dummy variables for the outliers. However, there is often a trade-off between an attempt to include all the relevant information in the model and a necessity to control its size.

Our time series of data consists of 24 observations and seven candidate explanatory variables. If we add five dummies for the potential outliers, the set of candidate explanatory variables can increase to as many as 12 variables. There is a high risk of obtaining heavily distorted regression results when the number of parameters to estimate is relatively large as compared to the number of observations. Here, after adding five dummies for outliers, there would be 13 estimated parameters (12 explanatory variables plus an intercept) and would be larger than half of the sample size. Thus, to protect our model from a dangerous overextension, we deliberately add to its set of candidate explanatory variables only one dummy variable, relating to the fourth quarter of 2007, the period for which Nokia reported outlying quarterly sales revenues. Our dummy variable, denoted as $IVq2007$, has the following value:

$$IVq2007 = \begin{cases} 1 & \text{in the 4th quarter of 2007} \\ 0 & \text{otherwise} \end{cases}$$

8.3.2   The "Two-Sigma Range" Applied to the Residuals from the General Regression

To check whether there are any significant multivariate outliers we ran the general regression, in which all the candidate explanatory variables are present, including a

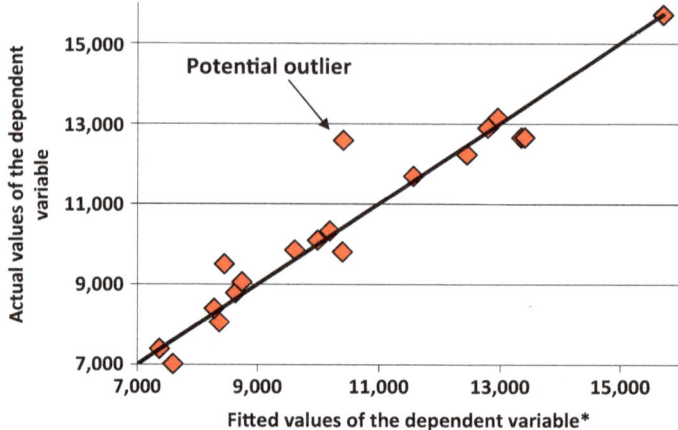

Fig. 8.1 Actual values of the dependent variable against the fitted values of the dependent variable from the general regression of Nokia's quarterly sales revenues. *From the general regression. Source: Nokia's quarterly financial statements; authors

dummy variable for the fourth quarter of 2007 (denoted as $IVq2007$). Then we scrutinized the obtained residuals for the presence of any outliers.

After regressing SR_t against its four lags, three seasonal dummy variables, and a dummy for the fourth quarter of 2007, we obtained the following model:

$$SR_t = 2.299,01 + 1,12\ SR_{t-1} + 0,02\ SR_{t-2} - 0,30\ SR_{t-3} + 0,06\ SR_{t-4}$$
$$- 3.614,49\ First - 706,96\ Second - 897,65\ Third + 990,73\ IVq2007$$

Then, for the obtained general regression, we computed the fitted values of the dependent variable and plotted them against the actual values of the dependent variable. This is shown in Fig. 8.1.

The visual inspection of Fig. 8.1 suggests that almost all observations lie near the core relationship. However, one observation seems to be a candidate for a multivariate outlier. Although it does not seem to be extremely outstanding, it lies significantly above the line which captures the core relationship. This observation should be taken into consideration and will be identified below.

The residuals obtained from the general regression of Nokia's quarterly sales revenues are presented in the following bar chart (Fig. 8.2).

The visual inspection of the above bar chart suggests that there are at least two "suspicious" residuals. They relate to the 1st quarter of 2006 (with residual equaling 1.061,22) and the 2nd quarter of 2007 (with residual equaling 2.171,78), respectively. Some other observations (e.g., the 2nd quarter of 2004) could perhaps also be considered to be outliers; however, for further analysis we will treat only the 1st quarter of 2006 and the 2nd quarter of 2007 as the potential outliers.

To formally test if these two "suspicious" observations are true multivariate outliers, we apply the "two-sigma range" to the obtained residuals. Similarly as in

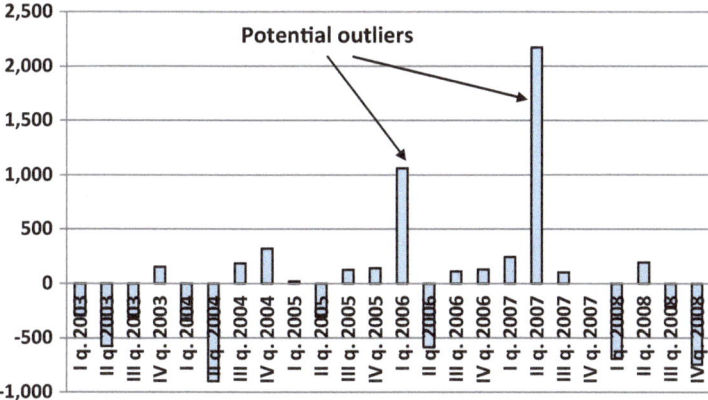

Fig. 8.2 The residuals* obtained from the general regression of Nokia's quarterly sales revenues. Source: Nokia's quarterly financial statements; authors

the case of individual variables, we set the upper and lower bounds on the ground of the respective arithmetic means and two standard deviations. Table 8.4 presents the results of the analysis.

Table 8.4 shows that in the case of only one observation (the 2nd quarter of 2007), the residual obtained from the general regression falls outside the range determined by the "two-sigma." Thus, we add to the set of candidate explanatory variables only one additional dummy, relating to the second quarter of 2007. This variable, denoted as $IIq2007$, has the following value:

$$IIq2007 = \begin{cases} 1 & \text{in the 2nd quarter of 2007} \\ 0 & \text{otherwise} \end{cases}$$

Before we proceed further, our updated dataset looks as presented in Table 8.5.

8.4 Selection of Explanatory Variables (from the Set of Candidates)

For our general model, which includes all the variables presented in Table 8.6, we obtained the following results.

Given the results shown in the above table, we should remove SR_{t-4} from our set of explanatory variables since it has an absolute value of its t-Statistic below the critical value and the smallest among all the explanatory variables. After removal of SR_{t-4}, we reestimate the parameters of the reduced model and repeat the procedure of the statistical significance analysis. The results of this re-estimation are presented below (Table 8.7).

Table 8.4 "Two-sigma range" applied to the residuals obtained from the general regression of Nokia's quarterly sales revenues

Period	Residual from the general regression	Is the observation an outlier according to the two-sigma range?
I q. 2003	−295,72	
II q. 2003	−577,07	
III q. 2003	−311,81	
IV q. 2003	154,27	
I q. 2004	−338,27	
II q. 2004	−898,13	
III q. 2004	184,79	
IV q. 2004	321,18	
I q. 2005	21,95	
II q. 2005	−304,24	
III q. 2005	125,58	
IV q. 2005	141,46	
I q. 2006	1.061,22	
II q. 2006	−585,59	
III q. 2006	113,00	
IV q. 2006	129,78	
I q. 2007	243,39	
II q. 2007	2.171,78	YES
III q. 2007	103,60	
IV q. 2007	0,00	
I q. 2008	−692,58	
II q. 2008	193,25	
III q. 2008	−215,16	
IV q. 2008	−746,68	
Arithmetic mean (AM)	0,00	
Standard deviation (SD)	630,21	
Upper bound (AM + 2 × SD)	1.260,42	
Lower bound (AM − 2 × SD)	−1.260,42	

Source: Nokia's quarterly financial statements; authors

We should remove now $IVq2007$ from the set of explanatory variables, because its absolute value of the t-Statistic is below the critical value and is the lowest among all the explanatory variables. We again reestimate the parameters of the reduced model after removal of $IVq2007$ and repeat the statistical significance analysis. The results of this second re-estimation are presented below (Table 8.8).

After removing *Third* from the set of explanatory variables, we again reestimate the parameters of this further reduced model. The results of this third re-estimation are presented below (Table 8.9).

We can see that all six remaining explanatory variables are statistically significant at the 1% significance level, meaning that we completed the procedure of variable selection. We began with a general model which included nine candidate

Table 8.5 The final dataset for the general regression of Nokia's quarterly sales revenues

Period	Dependent	Candidate explanatory variables								
	SR_t	SR_{t-1}	SR_{t-2}	SR_{t-3}	SR_{t-4}	First	Second	Third	IVq2007	IIq2007
I q. 2003	6.773	8.843	7.224	6.935	7.014	1	0	0	0	0
II q. 2003	7.019	6.773	8.843	7.224	6.935	0	1	0	0	0
III q. 2003	6.874	7.019	6.773	8.843	7.224	0	0	1	0	0
IV q. 2003	8.789	6.874	7.019	6.773	8.843	0	0	0	0	0
I q. 2004	6.625	8.789	6.874	7.019	6.773	1	0	0	0	0
II q. 2004	6.640	6.625	8.789	6.874	7.019	0	1	0	0	0
III q. 2004	6.939	6.640	6.625	8.789	6.874	0	0	1	0	0
IV q. 2004	9.063	6.939	6.640	6.625	8.789	0	0	0	0	0
I q. 2005	7.396	9.063	6.939	6.640	6.625	1	0	0	0	0
II q. 2005	8.059	7.396	9.063	6.939	6.640	0	1	0	0	0
III q. 2005	8.403	8.059	7.396	9.063	6.939	0	0	1	0	0
IV q. 2005	10.333	8.403	8.059	7.396	9.063	0	0	0	0	0
I q. 2006	9.507	10.333	8.403	8.059	7.396	1	0	0	0	0
II q. 2006	9.813	9.507	10.333	8.403	8.059	0	1	0	0	0
III q. 2006	10.100	9.813	9.507	10.333	8.403	0	0	1	0	0
IV q. 2006	11.701	10.100	9.813	9.507	10.333	0	0	0	0	0
I q. 2007	9.856	11.701	10.100	9.813	9.507	1	0	0	0	0
II q. 2007	12.587	9.856	11.701	10.100	9.813	0	1	0	0	1
III q. 2007	12.898	12.587	9.856	11.701	10.100	0	0	1	0	0
IV q. 2007	15.717	12.898	12.587	11.701	11.701	0	0	0	1	0
I q. 2008	12.660	15.717	12.898	12.587	9.856	1	0	0	0	0
II q. 2008	13.151	12.660	15.717	12.898	12.587	0	1	0	0	0
III q. 2008	12.237	13.151	12.660	15.717	12.898	0	0	1	0	0
IV q. 2008	12.662	12.237	13.151	12.660	15.717	0	0	0	0	0

Source: Nokia's quarterly financial statements; authors

Table 8.6 Regression results for the general model (i.e., model with the full set of all the candidate explanatory variables)

Variable	Coefficient	t-Statistic[b]
Intercept	2.677,68	3,919
SR_{t-1}	1,09	6,742
SR_{t-2}	0,33	1,174
SR_{t-3}	−0,54	−1,943
SR_{t-4}	−0,01	−0,049
First	−3.698,42	−6,808
Second	−1.840,35	−2,374
Third	−430,65	−0,510
IVq2007	385,73	0,497
IIq2007	2.802,25	5,286

R-squared: 0,98
Critical value of t-Statistic[a]: 2,977

Conclusion: remove SR_{t-4}

Source: Nokia's quarterly financial statements; authors.
[a]For 24 observations, 9 explanatory variables, and 1% significance level
[b]The smallest absolute value is noted

Table 8.7 Regression results for the reestimated model (i.e., after removing SR_{t-4} from the set of explanatory variables)

Variable	Coefficient	t-Statistic[b]
Intercept	2.651,84	6,277
SR_{t-1}	1,09	7,464
SR_{t-2}	0,32	1,249
SR_{t-3}	−0,55	−2,338
First	−3.681,15	−9,188
Second	−1.814,03	−3,345
Third	−402,47	−0,671
IVq2007	386,80	0,516
IIq2007	2.800,34	5,482

R-squared: 0,98
Critical value of t-Statistic[a]: 2,947

Conclusion: remove IVq2007

Source: Nokia's quarterly financial statements; authors
[a]For 24 observations, 8 explanatory variables, and 1% significance level
[b]The smallest absolute value is noted

explanatory variables and ended up with a specific model including six statistically significant explanatory variables.

Finally, the linear regression model of Nokia's quarterly sales revenues, obtained from the "general to specific modeling" procedure, looks as follows:

$$SR_t = 2.600,83 + 1,09\ SR_{t-1} + 0,47\ SR_{t-2} - 0,70\ SR_{t-3} - 3.636,03\ First$$
$$- 2.050,53\ Second + 2.876,59\ IIq2007$$

Table 8.8 Regression results for the reestimated model (i.e., after removing $IVq2007$ from the set of explanatory variables)

Variable	Coefficient	t-Statistic[b]
Intercept	2.634,46	6,404
SR_{t-1}	1,10	7,830
SR_{t-2}	0,40	1,940
SR_{t-3}	−0,64	−3,993
First	−3.723,67	−9,722
Second	−1.972,81	−4,522
Third	−240,63	*−0,482*
$IIq2007$	2.838,83	5,751

R-squared: 0,98
Critical value of t-Statistic[a]: 2,921

Conclusion: remove *Third*

Source: Nokia's quarterly financial statements; authors
[a]For 24 observations, 7 explanatory variables, and 1% significance level
[b]The smallest absolute value is noted

Table 8.9 Regression results for the reestimated model (i.e., after removing *Third* from the set of explanatory variables)

Variable	Coefficient	t-Statistic[b]
Intercept	2.600,83	6,565
SR_{t-1}	1,09	8,085
SR_{t-2}	0,47	*3,503*
SR_{t-3}	−0,70	−8,200
First	−3.636,03	−11,041
Second	−2.050,53	−5,177
$IIq2007$	2.876,59	6,041

R-squared: 0,98
Critical value of t-Statistic[a]: 2,898

Conclusion: all the explanatory variables are statistically significant at 1% significance level

Source: Nokia's quarterly financial statements; authors
[a]For 24 observations, 6 explanatory variables, and 1% significance level
[b]The smallest absolute value is noted

The model's R-squared equals 0,98 which means that it is capable of capturing the relationship between the Nokia's quarterly sales revenues (between the first quarter of 2003 and the fourth quarter of 2008) and the set of its explanatory variables. Such a high coefficient of determination suggests also a high predictive power of the obtained model (which will be tested later in the chapter). However, before the model is applied in practice, it must be verified for its statistical correctness. We will conduct such a verification in the following section.

It is also worth noting that only one out of two dummies for outliers turned out to be statistically significant. This is the multivariate outlier which was detected by means of "two-sigma" applied to the residuals from the general regression (and which was overlooked by "two-sigma" applied to the individual variables). It corroborates that the outlier detection approaches based on individual variables only may be misleading. Thus, it is always advisable to also search for the multivariate outliers.

8.5   Verification of the Obtained Model

We now apply the comprehensive procedure for verification of the resulting model, which will include the following tests:

- F-test for the general statistical significance of the whole model.
- Hellwig test for normality of distribution of residuals.
- F-test for autocorrelation of residuals.
- *ARCH-LM* test for heteroscedasticity of residuals.
- t-Student test for symmetry of residuals.
- Maximum series length test for randomness of residuals.
- Ramsey's *RESET* test for the general specification of the model.
- Variance inflation factor test for multicollinearity of explanatory variables.

8.5.1   F-test for the General Statistical Significance of the Model

Our model was estimated on the basis of $n = 24$ observations; there are $k = 6$ explanatory variables and the coefficient of determination, $R^2 = 0,980$. We then calculate the empirical F statistic with the following results:

$$F = \frac{R^2}{1 - R^2} \frac{n - (k + 1)}{k} = \frac{0,980}{1 - 0,980} \frac{24 - (6 + 1)}{6} = 138, 83.$$

For $n = 24$ and $k = 6$, we obtain:

$$m_1 = k = 6,$$
$$m_2 = n - (k + 1) = 24 - (6 + 1) = 17.$$

For the level of statistical significance set at $\alpha = 0,05$, and computed values of m_1 and m_2, we obtain the critical value of F^* from the Fisher–Snedecor table or from the Excel function = F.INV.RT(0.05,6,17), which equals 2,70.

We then compare the value of empirical statistic F with its critical value F^*. In our case $F > F^*$, which means that at least one slope parameter in our model is statistically different than zero.

8.5.2 Hellwig Test for Normality of Distribution of Residuals

To check the normality of residuals' distribution in our model by means of the Hellwig test, the first thing is to compute those residuals. This is presented in Table 8.10.

According to the Hellwig test procedure, we need to sort all the residuals in order of their increasing values and then standardize the residuals on the basis of their arithmetic average and their standard deviation. The next step is to obtain the normal distribution frequency from the table of normal distribution (or from Excel). We show the results of these three steps in the following table (Table 8.11).

In the next step, we create $n = 24$ cells, by dividing the range [0, 1] into n equal subranges. For $n = 24$, the individual cell width is $0,042 = 1/24$. Then, on the basis of the normal distribution frequencies appointed for the standardized residuals, we

Table 8.10 Computation of residuals from the model for Nokia's quarterly sales revenues

Period	SR_t (actual values of dependent variable)	$\widehat{SR_t}$ (fitted values of dependent variable)[a]	Residuals $(e_t = SR_t - \widehat{SR_t})$
I q. 2003	6.773	7.127,13	−354,13
II q. 2003	7.019	7.025,29	−6,29
III q. 2003	6.874	7.232,76	−358,76
IV q. 2003	8.789	8.640,47	148,53
I q. 2004	6.625	6.844,47	−219,47
II q. 2004	6.640	7.084,03	−444,03
III q. 2004	6.939	6.788,92	150,08
IV q. 2004	9.063	8.635,84	427,16
I q. 2005	7.396	7.438,18	−42,18
II q. 2005	8.059	8.005,54	53,46
III q. 2005	8.403	8.502,73	−99,73
IV q. 2005	10.333	10.356,43	−23,43
I q. 2006	9.507	8.515,57	991,43
II q. 2006	9.813	9.873,63	−60,63
III q. 2006	10.100	10.515,65	−415,65
IV q. 2006	11.701	11.550,16	150,84
I q. 2007	9.856	9.574,87	281,13
II q. 2007	12.587	12.587,00	0,00
III q. 2007	12.898	12.736,65	161,35
IV q. 2007	15.717	15.655,08	61,92
I q. 2008	12.660	13.316,79	−656,79
II q. 2008	13.151	12.693,50	457,50
III q. 2008	12.237	11.861,25	375,75
IV q. 2008	12.662	13.240,04	−578,04

Source: Nokia's quarterly financial statements; authors
[a]Computed according to the obtained model, $SR_t = 2.600,83 + 1,09\ SR_{t-1} + 0,47\ SR_{t-2} - 0,70\ SR_{t-3} - 3.636,03\ First - 2.050,53\ Second + 2.876,59\ IIq2007$

Table 8.11 The results of the residual sorting, standardization, and derivation of the normal distribution frequencies

Residuals $(e_t = SR_t - \widehat{SR}_t)$	Sorted residuals e_{ts}	Standardized residuals e'_{ts}	Normal distribution frequency for standardized residuals $F(e_{ts})^a$
−354,13	−656,79	−1,775	0,038
−6,29	−578,04	−1,562	0,059
−358,76	−444,03	−1,200	0,115
148,53	−415,65	−1,123	0,131
−219,47	−358,76	−0,970	0,166
−444,03	−354,13	−0,957	0,169
150,08	−219,47	−0,593	0,277
427,16	−99,73	−0,270	0,394
−42,18	−60,63	−0,164	0,435
53,46	−42,18	−0,114	0,455
−99,73	−23,43	−0,063	0,475
−23,43	−6,29	−0,017	0,493
991,43	0,00	0,000	0,500
−60,63	53,46	0,144	0,557
−415,65	61,92	0,167	0,566
150,84	148,53	0,401	0,656
281,13	150,08	0,406	0,657
0,00	150,84	0,408	0,658
161,35	161,35	0,436	0,669
61,92	281,13	0,760	0,776
−656,79	375,75	1,016	0,845
457,50	427,16	1,154	0,876
375,75	457,50	1,236	0,892
−578,04	991,43	2,680	0,996
Arithmetic mean ($\bar{e}$)	0	0	–
Std. deviation (S)	370,00	1	–

Source: Nokia's quarterly financial statements; authors
[a]The values derived from the table for normal distribution frequency or from the spreadsheet

compute the number of residuals falling into each individual cell. Finally, we compute the value of k_e, which is the number of empty cells. All these three steps are presented in Table 8.12.

As the above data show, the number of empty cells $k_e = 8$. Then we obtain the critical Hellwig test values of k_1 and k_2, for $n = 24$ observations and for the assumed level of statistical significance α (we assume $\alpha = 0,05$). In the appropriate table (found in the Appendix A4), we find that $k_1 = 5$ and $k_2 = 11$.

Since $k_1 \leq k_e \leq k_2$, we conclude that no statistical evidence was found against the normality of the residuals from our model of the Nokia's quarterly sales revenues.

Table 8.12 The computation of the number of empty cells for Hellwig test

Cell	Number of residuals falling into the cell[a]
<0,000; 0,042)	1
<0,042; 0,083)	1
<0,083; 0,125)	1
<0,125; 0,167)	2
<0,167; 0,209)	1
<0,209; 0,250)	0
<0,250; 0,292)	1
<0,292; 0,334)	0
<0,334; 0,375)	0
<0,375; 0,417)	1
<0,417; 0,459)	2
<0,459; 0,500)	3
<0,500; 0,542)	0
<0,542; 0,584)	2
<0,584; 0,626)	0
<0,626; 0,667)	3
<0,667; 0,709)	1
<0,709; 0,751)	0
<0,751; 0,792)	1
<0,792; 0,834)	0
<0,834; 0,876)	1
<0,876; 0,917)	2
<0,917; 0,959)	0
<0,959; 1,000>	1
The number of empty cells (k_e)	8

Source: Nokia's quarterly financial statements; authors
[a]On the basis of the data from the last column of Table 8.11

8.5.3 F-test for the Autocorrelation of Residuals

For checking the presence of residuals autocorrelation by means of the F-test, we arbitrarily consider the autocorrelation of up to fourth quarter. The data prepared for the regression of residuals against lagged residuals are shown below (Table 8.13).

We now regress the residuals against the lagged residuals. The results of this regression are presented in Table 8.14.

Although our model, which is tested for autocorrelation, was estimated on the basis of 24 observations, in testing the autocorrelation of up to fourth order we have only 20 observations, because we lost 4 observations by lagging residuals. This means that for the F-test for autocorrelation $n = 20$, there are four explanatory variables (which means that $k = 4$) and the coefficient of determination $R^2 = 0,344$. Calculating the empirical statistic, F produces the following results:

Table 8.13 The residuals and lagged residuals up to the fourth order from our model of the Nokia's quarterly sales revenues

Residuals e_t	First lag e_{t-1}	Second lag e_{t-2}	Third lag e_{t-3}	Fourth lag e_{t-4}
−354,13	–	–	–	–
−6,29	−354,13	–	–	–
−358,76	−6,29	−354,13	–	–
148,53	−358,76	−6,29	−354,13	–
−219,47	148,53	−358,76	−6,29	−354,13
−444,03	−219,47	148,53	−358,76	−6,29
150,08	−444,03	−219,47	148,53	−358,76
427,16	150,08	−444,03	−219,47	148,53
−42,18	427,16	150,08	−444,03	−219,47
53,46	−42,18	427,16	150,08	−444,03
−99,73	53,46	−42,18	427,16	150,08
−23,43	−99,73	53,46	−42,18	427,16
991,43	−23,43	−99,73	53,46	−42,18
−60,63	991,43	−23,43	−99,73	53,46
−415,65	−60,63	991,43	−23,43	−99,73
150,84	−415,65	−60,63	991,43	−23,43
281,13	150,84	−415,65	−60,63	991,43
0,00	281,13	150,84	−415,65	−60,63
161,35	0,00	281,13	150,84	−415,65
61,92	161,35	0,00	281,13	150,84
−656,79	61,92	161,35	0,00	281,13
457,50	−656,79	61,92	161,35	0,00
375,75	457,50	−656,79	61,92	161,35
−578,04	375,75	457,50	−656,79	61,92

Source: Nokia's quarterly financial statements; authors

Table 8.14 Results of regressing residuals against lagged (up to fourth order) residuals

Variable	Coefficient	t-Statistic
Intercept	57,54	0,703
First lag e_{t-1}	−0,17	−0,639
Second lag e_{t-2}	−0,58	−2,370
Third lag e_{t-3}	0,15	0,544
Fourth lag e_{t-4}	−0,13	−0,491
R-squared: 0,344		

Source: Nokia's quarterly financial statements; authors

$$F = \frac{R^2}{1 - R^2} \frac{n - (k + 1)}{k} = \frac{0,344}{1 - 0,344} \frac{20 - (4 + 1)}{4} = 1,97.$$

Now we need to obtain the critical value of F^*. For $n = 20$ and $k = 4$, we obtain:

$$m_1 = k = 4,$$
$$m_2 = n - (k + 1) = 20 - (4 + 1) = 15.$$

For the level of statistical significance set at $\alpha = 0,05$, and computed values of m_1 and m_2, we obtain the critical value of F^* from the Fisher–Snedecor table, which equals 3,06.

Finally, we compare the value of the empirical statistic F with its critical value F^*. In our case $F < F^*$, which means that our model for the Nokia's quarterly sales revenues is free from any serial autocorrelation of the residuals.

8.5.4 ARCH-LM F-test for Heteroscedasticity of Residuals

In checking the presence of residuals' heteroscedasticity in our model by means of the *ARCH-LM* test, the first thing to do is to prepare the data by squaring and lagging our residuals to the specified maximum considered lag (k). Similarly as when testing autocorrelation, we consider the lags of up to fourth order. The data prepared for the regression of squared residuals against lagged squared residuals are shown below (Table 8.15).

The results of the regression of squared residuals against lagged squared residuals are presented in Table 8.16.

For our *ARCH-LM* test for heteroscedasticity $n = 20$, there are four explanatory variables ($k = 4$) and the coefficient of determination $R^2 = 0,094$. Computing the empirical statistic F produces the following results:

$$F = \frac{R^2}{1 - R^2} \frac{n - (k + 1)}{k} = \frac{0,094}{1 - 0,094} \frac{20 - (4 + 1)}{4} = 0,39.$$

Now we obtain the critical value of F^*. For $n = 20$ and $k = 4$,

$$m_1 = k = 4,$$
$$m_2 = n - (k + 1) = 20 - (4 + 1) = 15.$$

For the level of statistical significance set at $\alpha = 0,05$, and computed values of m_1 and m_2, the critical F^* from the Fisher–Snedecor table equals 3,06.

Finally, we need to compare the value of empirical statistic F with its critical value F^*. In our case, $F < F^*$. Thus, we conclude that there is not any significant heteroscedasticity of the residuals in our model for the Nokia's quarterly sales revenues.

Table 8.15 The residuals, squared residuals, and lagged squared residuals from our model of the Nokia's quarterly sales revenues

Residuals e_t	Squared residuals e_t^2	First lag of squared residuals e_{t-1}^2	Second lag of squared residuals e_{t-2}^2	Third lag of squared residuals e_{t-3}^2	Fourth lag of squared residuals e_{t-4}^2
−354,13	125.408,06	–	–	–	–
−6,29	39,56	125.408,06	–	–	–
−358,76	128.708,74	39,56	125.408,06	–	–
148,53	22.061,16	128.708,74	39,56	125.408,06	–
−219,47	48.167,08	22.061,16	128.708,74	39,56	125.408,06
−444,03	197.162,64	48.167,08	22.061,16	128.708,74	39,56
150,08	22.524,01	197.162,64	48.167,08	22.061,16	128.708,74
427,16	182.465,67	22.524,01	197.162,64	48.167,08	22.061,16
−42,18	1.779,15	182.465,67	22.524,01	197.162,64	48.167,08
53,46	2.857,97	1.779,15	182.465,67	22.524,01	197.162,64
−99,73	9.946,07	2.857,97	1.779,15	182.465,67	22.524,01
−23,43	548,96	9.946,07	2.857,97	1.779,15	182.465,67
991,43	982.933,44	548,96	9.946,07	2.857,97	1.779,15
−60,63	3.676,00	982.933,44	548,96	9.946,07	2.857,97
−415,65	172.764,92	3.676,00	982.933,44	548,96	9.946,07
150,84	22.752,71	172.764,92	3.676,00	982.933,44	548,96
281,13	79.034,08	22.752,71	172.764,92	3.676,00	982.933,44
0,00	0,00	79.034,08	22.752,71	172.764,92	3.676,00
161,35	26.033,82	0,00	79.034,08	22.752,71	172.764,92
61,92	3.834,09	26.033,82	0,00	79.034,08	22.752,71
−656,79	431.373,10	3.834,09	26.033,82	0,00	79.034,08
457,50	209.306,25	431.373,10	3.834,09	26.033,82	0,00
375,75	141.188,06	209.306,25	431.373,10	3.834,09	26.033,82
−578,04	334.130,24	141.188,06	209.306,25	431.373,10	3.834,09

Source: Nokia's quarterly financial statements; authors

Table 8.16 Results of regressing squared residuals against lagged (up to fourth order) squared residuals

Variable	Coefficient	t-Statistic
Intercept	224.130,41	2,407
First lag e_{t-1}^2	−0,22	−0,857
Second lag e_{t-2}^2	−0,04	−0,174
Third lag e_{t-3}^2	−0,18	−0,684
Fourth lag e_{t-4}^2	−0,25	−0,928
R-squared: 0,094		

Source: Nokia's quarterly financial statements; authors

8.5.5 *t-Student Test for Symmetry of Residuals*

In checking the symmetry of residuals in our model, the first thing to do is to compute the number of positive residuals (denoted as m_{e1}) and the number of negative residuals (denoted as m_{e2}). This is shown in Table 8.17.

Then we calculate the empirical statistics as follows:

$$t_{e1} = \frac{\left| \frac{m_{e1}}{n} - \frac{1}{2} \right|}{\sqrt{\frac{\frac{m_{e1}}{n}\left(1 - \frac{m_{e1}}{n}\right)}{n-1}}} = \frac{\left| \frac{11}{24} - \frac{1}{2} \right|}{\sqrt{\frac{\frac{11}{24}\left(1 - \frac{11}{24}\right)}{24-1}}} = 0,401,$$

Table 8.17 The computations for the symmetry test (for symmetry of residuals from our model of the Nokia's quarterly sales revenues)

Residuals ($e_t = SR_t - \widehat{SR}_t$)	Sign of residual (1 – positive, 0 – otherwise)	Sign of residual (1 – negative, 0 – otherwise)
−354,13	0	1
−6,29	0	1
−358,76	0	1
148,53	1	0
−219,47	0	1
−444,03	0	1
150,08	1	0
427,16	1	0
−42,18	0	1
53,46	1	0
−99,73	0	1
−23,43	0	1
991,43	1	0
−60,63	0	1
−415,65	0	1
150,84	1	0
281,13	1	0
0,00	0	0
161,35	1	0
61,92	1	0
−656,79	0	1
457,50	1	0
375,75	1	0
−578,04	0	1
$m_{e1}=$	11	–
$m_{e2}=$	–	12

Source: Nokia's quarterly financial statements; authors

$$t_{e2} = \frac{\left|\frac{m_{e1}}{n} - \frac{1}{2}\right|}{\sqrt{\frac{\frac{m_{e1}}{n}\left(1-\frac{m_{e1}}{n}\right)}{n-1}}} = \frac{\left|\frac{12}{24} - \frac{1}{2}\right|}{\sqrt{\frac{\frac{12}{24}\left(1-\frac{12}{24}\right)}{24-1}}} = 0,000.$$

The next step is to obtain the critical value of t_t, from the t-Student distribution table or from Excel for $n - 1 = 23$ degrees of freedom and for the assumed level of statistical significance $\alpha = 5\%$ (= T.INV(0.05,23)). This value of t_t equals 2,069. Given that in our model $t_{e1} < t_t$ as well as $t_{e2} < t_t$, we may conclude that no statistical evidence was found against the symmetry of the residuals from our model of the Nokia's quarterly sales revenues.

8.5.6 Maximum Series Length Test for Randomness of Residuals

For the maximum series length test, we denote the positive residuals as "A" and negative residuals as "B." Then we compute the length of the longest series of residuals of the same sign (k_e). This is shown in the following table (Table 8.18).

We then obtain the critical value of k_t from the maximum series length table (included in the Appendix A6). For 24 residuals ($n = 24$) and for the assumed level of statistical significance α of 5%, we find that $k_t = 7$. Given that in our model $k_e < k_t$, we may conclude that no statistical evidence was found against the randomness of the residuals from our model of the Nokia's quarterly sales revenues.

8.5.7 Ramsey's RESET Test for the General Specification of the Model

In applying the Ramsey's *RESET* test, the first thing to do is to prepare the data by computing the fitted values of the dependent variable as well as the values of auxiliary explanatory variables. We will add two auxiliary variables:

- Squared fitted values of the dependent variable obtained from our model (denoted as *FITTED*^{2}).
- Cubic fitted values of the dependent variable obtained from our model (denoted as *FITTED*^{3}).

These data are presented in Table 8.19. Table 8.20, in turn, presents the results of auxiliary regression estimated for Ramsey's RESET test for our model of the Nokia's quarterly sales revenues.

Table 8.18 The computations for the maximum series length test (for randomness of residuals from our model of the Nokia's quarterly sales revenues)

Residuals ($e_t = SR_t - \widehat{SR}_t$)	Sign of residual ("A" – positive, "B" – negative)	Length of the series of residuals of the same sign
−354,13	B	
−6,29	B	
−358,76	B	3
148,53	A	1
−219,47	B	
−444,03	B	2
150,08	A	
427,16	A	2
−42,18	B	1
53,46	A	1
−99,73	B	
−23,43	B	2
991,43	A	1
−60,63	B	
−415,65	B	2
150,84	A	
281,13	A	2
0,00	−	
161,35	A	
61,92	A	2
−656,79	B	1
457,50	A	
375,75	A	2
−578,04	B	1
Length of the longest series of residuals (k_e)		3

Source: Nokia's quarterly financial statements; authors

We know from the F-test for the general significance that the R-squared of our model equals 0,980. Table 8.20, in turn, informs us that the R-squared of the auxiliary regression equals 0,984. Our tested model was estimated on the basis of 24 observations ($n = 24$) and it includes six explanatory variables (which means that $k = 6$). Therefore, our empirical F statistic is computed as follows:

$$F = \frac{\left(R_2^2 - R_1^2\right)/2}{\left(1 - R_2^2\right)/(n - (k+3))} = \frac{(0,984 - 0,980)/2}{(1 - 0,984)/(24 - (6+3))} = 1,88.$$

Table 8.19 The data for Ramsey's *RESET* test for correctness of specification of our model of the Nokia's quarterly sales revenues

Period	$\widehat{SR_t}$ (fitted values of dependent variable)	$FITTED^2$	$FITTED^3$
I q. 2003	7.127,13	50.795.932,46	362.029.037.439,35
II q. 2003	7.025,29	49.354.762,25	346.731.737.785,25
III q. 2003	7.232,76	52.312.753,34	378.365.358.816,34
IV q. 2003	8.640,47	74.657.781,22	645.078.575.536,73
I q. 2004	6.844,47	46.846.739,61	320.641.001.313,52
II q. 2004	7.084,03	50.183.524,19	355.501.743.727,57
III q. 2004	6.788,92	46.089.474,13	312.897.886.330,32
IV q. 2004	8.635,84	74.577.769,47	644.041.844.271,34
I q. 2005	7.438,18	55.326.514,00	411.528.541.194,63
II q. 2005	8.005,54	64.088.731,18	513.065.143.090,49
III q. 2005	8.502,73	72.296.500,91	614.717.981.955,81
IV q. 2005	10.356,43	107.255.584,18	1.110.784.648.422,54
I q. 2006	8.515,57	72.514.872,76	617.505.221.029,07
II q. 2006	9.873,63	97.488.562,80	962.565.965.841,70
III q. 2006	10.515,65	110.578.992,40	1.162.810.493.975,46
IV q. 2006	11.550,16	133.406.309,17	1.540.864.869.283,56
I q. 2007	9.574,87	91.678.101,47	877.805.740.430,15
II q. 2007	12.587,00	158.432.569,00	1.994.190.746.003,00
III q. 2007	12.736,65	162.222.201,75	2.066.167.078.098,85
IV q. 2007	15.655,08	245.081.480,48	3.836.769.797.377,33
I q. 2008	13.316,79	177.336.934,97	2.361.558.982.432,36
II q. 2008	12.693,50	161.124.918,13	2.045.238.995.174,64
III q. 2008	11.861,25	140.689.320,78	1.668.751.616.670,44
IV q. 2008	13.240,04	175.298.791,22	2.320.963.881.733,21

Source: Nokia's quarterly financial statements; authors

Table 8.20 Auxiliary regression estimated for Ramsey's *RESET* test conducted for our model of the Nokia's quarterly sales revenues

Variable	Coefficient	t-Statistic
Intercept	−1.415,51	−0,669
SR_{t-1}	5,68	2,211
SR_{t-2}	2,48	2,224
SR_{t-3}	−3,61	−2,240
First	−18.952,99	−2,216
Second	−10.678,08	−2,227
IIq2007	15.248,72	2,198
$FITTED^2$	0,00	−1,731
$FITTED^3$	0,00	1,664
R-squared: 0,984		

Source: Nokia's quarterly financial statements; authors

Now we need to obtain the critical value of F^*. For $n = 24$ and $k = 6$, we obtain:

$$m_1 = 2,$$
$$m_2 = n - (k + 3) = 24 - (6 + 3) = 15.$$

For the level of statistical significance set at $\alpha = 0,05$, and determined values of m_1 and m_2, we obtain the critical value of F^* from the Fisher–Snedecor table, which equals 3,68.

Finally, we compare the value of the empirical statistic F with its critical value F^*. In our case $F < F^*$, which means that none of the two auxiliary variables is statistically significant. Therefore, we can conclude that our model of the Nokia's quarterly sales revenues is correctly specified.

8.5.8   Variance Inflation Factor Test for the Multicollinearity of Explanatory Variables

Finally, we test the potential multicollinearity of explanatory variables in our model. It includes six explanatory variables, which means that we need to run six separate tests for multicollinearity. The results of the auxiliary regressions are presented in Table 8.21. As we can see, our model is not free from multicollinearity of the explanatory variables, because the empirical VIF_i statistics for all three lags of dependent variable have values above the critical threshold.

In light of the detected multicollinearity of the explanatory variables, we recommend caution when interpreting its individual slope coefficients as well as their respective t-Statistics. However, the multicollinearity tends to understate the t-Statistics and in our model they all exceed the critical value. Thus, it is unlikely that the distortion of t-Statistics resulting from multicollinearity may significantly affect our model. In contrast, it could significantly distort the values of the individual slope coefficients. However, our model is aimed at forecasting, which means that it may be useful even without the interpretation of its individual parameters, since it is capable of generating predictions with satisfactory accuracy.

Table 8.21 Results of variance inflation factor tests run for explanatory variables in our model of the Nokia's quarterly sales revenues

Auxiliary regression	R-squared	VIF_i statistic
SR_{t-1} against remaining five variables	0,931	14,46
SR_{t-2} against remaining five variables	0,930	14,37
SR_{t-3} against remaining five variables	0,815	5,42
First against remaining five variables	0,620	2,63
Second against remaining five variables	0,738	3,81
IIq2007 against remaining five variables	0,148	1,17
Critical value of VIF_i statistic (as a "rule of thumb"): 5,00		

Source: Nokia's quarterly financial statements; authors

8.6   Evaluation of the Predictive Power of the Estimated Model

In this section, we first construct one-quarter-ahead predictions of the Nokia's quarterly sales revenues and then check their individual and average forecast errors. Our model, which was verified for correctness in the preceding section, was built on the estimation subsample covering 24 quarters, from the first quarter of 2003 through the fourth quarter of 2008. However, our full data set (presented in Table 8.1) extends to the end of 2011. Thus, we have 12 more observations which will serve as our forecast subsample.

To evaluate the predictive capabilities of our model, we formulate 12 one-quarter-ahead forecasts. Our model is autoregressive, with the shortest lag of a dependent variable equaling one quarter. Thus, it enables predicting the Nokia's quarterly sales revenues for a given quarter based on the information available in the preceding quarter. The forecast of the dependent variable for the first quarter of 2009 (the first observation in our forecast subsample) may be done on the ground of our estimation subsample as well as the parameters of the model estimated for this subsample. However, in the case of the following quarters, it is legitimate to update the model parameters recursively, after adding one additional observation. Thus, our procedure of recursive forecast-building is as follows:

- The forecast for the first quarter of 2009 is based on the model estimated on the original estimation sample, covering the period between the first quarter of 2003 and the fourth quarter of 2008 (24 observations).
- The forecast for the second quarter of 2009 is based on the updated (re-estimated) model, covering the re-estimation subsample starting in the first quarter of 2003 and ending in the first quarter of 2009 (25 observations).
- We continue this same pattern in calculating the following forecasts until the last fourth quarter of 2011.
- The forecast for the fourth quarter of 2011 is based on the updated (reestimated) model, covering the re-estimation subsample starting in the first quarter of 2003 and ending in the third quarter of 2011 (35 observations).

Table 8.22 contains the results of the recursive re-estimations of our model as well as the obtained one-quarter-ahead forecasts of the Nokia's quarterly sales revenues.

Table 8.23 compares the Nokia's actual quarterly sales revenues with their one-quarter-ahead forecasts, generated from the estimated autoregressive model. The model seems to have quite good average predictive capability (with the median absolute percentage forecast error not exceeding 5%). With the exception of two quarters, the absolute percentage forecast errors are single digit. Thus, the model seems to be useful in making short-term forecasts of the Nokia's quarterly sales revenues (at least in 2009–2011), particularly in light of its simplicity. However, it cannot be deemed fully reliable, because there are periods for which it significantly understated or overstated the sales revenues (the second quarter of 2009 and the

Table 8.22 Results of the recursive re-estimations of our model as well as the obtained one-quarter-ahead forecasts of the Nokia's quarterly sales revenues

Forecast for	Estimation sample		Intercept	SR_{t-1}	SR_{t-2}	SR_{t-3}	First	Second	IIq2007	R-squared	Forecast obtained
I q. 2009	I q. 2003–IV q. 2008	Parameters[a]	2.600,83 (6,565)	1,09 (8,085)	0,47 (3,503)	−0,70 (−8,200)	−3.636,03 (−11,041)	−2.050,53 (−5,177)	2.876,59 (6,041)	0,980	9.290,69
		Observations	–	12.662	12.237	13.151	1	0	0		
II q. 2009	I q. 2003–I q. 2009	Parameters[a]	2.603,78 (6,980)	1,09 (9,238)	0,47 (3,757)	−0,70 (−9,033)	−3.641,42 (−13,563)	−2.048,41 (−5,407)	2.877,22 (6,223)	0,980	8.030,81
		Observations	–	9.274	12.662	12.237	0	1	0		
III q. 2009	I q. 2003–II q. 2009	Parameters[a]	2.435,80 (5,134)	0,94 (6,635)	0,54 (3,431)	−0,61 (−6,472)	−3.374,54 (−10,193)	−1.924,81 (−3,980)	2.581,75 (4,425)	0,965	9.105,74
		Observations	–	9.912	9.274	12.662	0	0	0		
IV q. 2009	I q. 2003–III q. 2009	Parameters[a]	2.413,70 (5,053)	0,95 (6,623)	0,50 (3,223)	−0,56 (−6,557)	−3.395,97 (−10,197)	−1.854,78 (−3,837)	2.567,51 (4,368)	0,963	11.404,53
		Observations	–	9.810	9.912	9.274	0	0	0		
I q. 2010	I q. 2003–IV q. 2009	Parameters[a]	2.469,69 (5,202)	0,93 (6,553)	0,53 (3,491)	−0,58 (−6,890)	−3.432,00 (−10,359)	−1.974,75 (−4,212)	2.575,50 (4,380)	0,962	9.649,13
		Observations	–	11.988	9.810	9.912	1	0	0		
II q. 2010	I q. 2003–I q. 2010	Parameters[a]	2.477,69 (5,352)	0,93 (6,714)	0,53 (3,602)	−0,58 (−7,050)	−3.443,33 (−10,758)	−1.982,85 (−4,337)	2.577,28 (4,481)	0,962	10.020,48
		Observations	–	9.522	11.988	9.810	0	1	0		
III q. 2010	I q. 2003–II q. 2010	Parameters[a]	2.478,39 (5,480)	0,93 (6,872)	0,53 (3,692)	−0,58 (−7,212)	−3.443,50 (−11,002)	−1.984,50 (−4,470)	2.579,75 (4,633)	0,962	9.875,44
		Observations	–	10.003	9.522	11.988	0	0	0		
IV q. 2010	I q. 2003–III q. 2010	Parameters[a]	2.475,34 (5,531)	0,93 (6,945)	0,52 (3,669)	−0,57 (−7,346)	−3.457,62 (−11,185)	−1.974,25 (−4,496)	2.574,68 (4,672)	0,961	11.817,44
		Observations	–	10.270	10.003	9.522	0	0	0		

(continued)

Table 8.22 (continued)

Forecast for	Estimation sample	–	Intercept	SR_{t-1}	SR_{t-2}	SR_{t-3}	First	Second	IIq2007	R-squared	Forecast obtained
I q. 2011	I q. 2003–IV q. 2010	Parameters[a]	2.520,71 (5,493)	0,93 (6,772)	0,54 (3,739)	−0,59 (−7,535)	−3.528,97 (−11,234)	−2.080,93 (−4,669)	2.577,93 (4,552)	0,959	10.445,59
		Observations	–	12.651	10.270	10.003	1	0	0		
II q. 2011	I q. 2003–I q. 2011	Parameters[a]	2.524,67 (5,641)	0,93 (6,976)	0,54 (3,838)	−0,59 (−7,685)	−3.531,25 (−11,507)	−2.084,20 (−4,787)	2.578,79 (4,644)	0,959	10.927,12
		Observations	–	10.399	12.651	10.270	0	1	0		
III q. 2011	I q. 2003–II q. 2011	Parameters[a]	2.659,01 (5,259)	0,92 (6,068)	0,53 (3,283)	−0,57 (−6,632)	−3.501,20 (−10,053)	−2.258,26 (−4,610)	2.792,50 (4,466)	0,945	9.373,11
		Observations	–	9.275	10.399	12.651	0	0	0		
IV q. 2011	I q. 2003–III q. 2011	Parameters[a]	2.647,74 (5,302)	0,95 (6,767)	0,51 (3,268)	−0,59 (−7,075)	−3.539,33 (−10,462)	−2.189,66 (−4,654)	2.784,90 (4,508)	0,945	9.758,75
		Observations	–	8.980	9.275	10.399	0	0	0		

Source: Nokia's quarterly financial statements; authors

[a]t-Statistics in parentheses

Table 8.23 Forecast errors of the Nokia's quarterly sales revenues

Period	Nokia's actual quarterly sales revenues[a]	Nokia's forecasted quarterly sales revenues[b]	Percentage forecast error (%)[c]	Absolute percentage forecast error (%)
I q. 2009	9.274	9.290,69	−0,18	0,18
II q. 2009	9.912	8.030,81	23,42	23,42
III q. 2009	9.810	9.105,74	7,73	7,73
IV q. 2009	11.988	11.404,53	5,12	5,12
I q. 2010	9.522	9.649,13	−1,32	1,32
II q. 2010	10.003	10.020,48	−0,17	0,17
III q. 2010	10.270	9.875,44	4,00	4,00
IV q. 2010	12.651	11.817,44	7,05	7,05
I q. 2011	10.399	10.445,59	−0,45	0,45
II q. 2011	9.275	10.927,12	−15,12	15,12
III q. 2011	8.980	9.373,11	−4,19	4,19
IV q. 2011	10.005	9.758,75	2,52	2,52
Arithmetic mean			2,37	5,94
Median			1,17	4,09

Source: Nokia's quarterly financial statements; authors
[a]Data from the last column of Table 8.1
[b]Data from the last column of Table 8.22
[c][(Actual sales revenues/forecasted sales revenues) − 1] × 100

second quarter of 2011, respectively). Thus, as is often the case, the model's predictions should be combined with other relevant information (e.g., macroeconomic data or qualitative information about the company's actions, such as the introduction of new products).

Financial statement users (such as stock market investors) often trace not only the monetary amounts of company's sales revenues but also their year-over-year growth rates. This is so because markets tend to expect repeatedly positive growth rates (particularly from high-tech companies) and to treat falling sales as a signal of the company's weaknesses (e.g., deteriorating competitive position). As a result, declining sales often entail falling stock prices. Thus, it is legitimate to check the extent to which our model is able to predict the turning points of the Nokia's sales growth rates.

In order to check the prediction of turning points, the actual and forecasted growth rates were computed and compared. For any quarter, the growth rates were computed as follows:

$$\text{Actual sales growth} = \left[\left(\frac{SR_t}{SR_{t-4}} \right) - 1 \right] \times 100,$$

Table 8.24 Actual and predicted year-over-year changes of the Nokia's quarterly sales revenues

Period	Nokia's actual change of quarterly sales revenues (%)	Nokia's forecasted change of quarterly sales revenues (%)
I q. 2009	−26,7	−26,6
II q. 2009	−24,6	−38,9
III q. 2009	−19,8	−25,6
IV q. 2009	−5,3	−9,9
I q. 2010	2,7	4,0
II q. 2010	0,9	1,1
III q. 2010	4,7	0,7
IV q. 2010	5,5	−1,4
I q. 2011	9,2	9,7
II q. 2011	−7,3	9,2
III q. 2011	−12,6	−8,7
IV q. 2011	−20,9	−22,9

Source: Nokia's quarterly financial statements; authors

$$\text{Forecasted sales growth} = \left[\left(\frac{\widehat{SR_t}}{SR_{t-4}} \right) - 1 \right] \times 100,$$

where:

SR_t—Nokia's actual quarterly sales revenues (in EUR millions) in t-th quarter,

$\widehat{SR_t}$—Nokia's forecasted quarterly sales revenues (in EUR millions) in t-th quarter.

Table 8.24 compares the actual and predicted growth rates of the Nokia's quarterly sales revenues. As might be seen, for majority of quarters the estimated model correctly predicted direction of year-over-year changes of the investigated variable. The most notable exception is the second quarter of 2011 for which the model predicted growing sales while they actually fell. It corroborates that our model is not fully reliable, and its use should be combined with some other relevant information.

It is also worth noting that the model was capable of warning against all the particularly deep declines of the Nokia's sales revenues. The model generated forecasts of negative growth rates for all five quarters when Nokia suffered from double-digit downturns of sales.

Chapter 9
Real-Life Case Study: Identifying Overvalued and Undervalued Airlines

9.1 Introduction

The preceding chapter offered a real-life example which illustrated a construction and application of a time-series econometric model. In contrast, in this chapter, we present an example of a cross-sectional model, based on the observed stock prices and financial results of selected airline companies, listed on global stock exchanges. The purpose of the model discussed in this chapter is to capture the statistical relationships between relative market values of selected airlines (as measured by their price-to-sales ratios) on one side and their selected accounting ratios, computed on the basis of their reported financial statements. Finally, the obtained model will be used in identifying potentially undervalued and overvalued stocks.

Similarly as for the model discussed in Chap. 8, we first propose a preliminary set of explanatory variables and check for the presence of any potential univariate and multivariate outliers in the dataset. Then we apply the procedure of "general to specific" modeling to select the final set of explanatory variables and test for its statistical correctness. Finally, we close the chapter with the evaluation of the model capability of picking undervalued and overvalued stocks.

9.2 Preliminary Specification of the Model

We are interested in building and testing a model which could be useful in identifying potentially overvalued and undervalued shares of airline companies. We define overvalued stocks as those whose actual stock prices exceed their true (fundamental) values while undervalued ones as those in which case the observed stock prices are lower than the estimated fundamental values. When building investment portfolios, it is often assumed that under- and overvalued stocks offer

J. Welc, P.J.R. Esquerdo, *Applied Regression Analysis for Business*,
https://doi.org/10.1007/978-3-319-71156-0_9

a high probability of providing above-average and below-average future stock returns, respectively.

Thus, our goal is to compare the per-share market values of individual airlines in search of investment opportunities. However, the stock prices of various firms are not directly comparable. Instead, we may compare the relative values of companies on the ground of so-called valuation multiples (ratios), such as price-to-earnings (P/E), price-to-book-value (P/BV), or price-to-sales (P/S) ratios. In our model, we will use price-to-sales ratio (later in the chapter denoted as P/S), because it enables comparing relative market values of almost all companies, including those with negative earnings or negative shareholder's equity.

In order to test a model's capability of identifying mispriced stocks, the model is estimated on the ground of the data available as at the end of June 2011. Then, three alternative stock investment horizons will be assumed: 1-year, 2-year, and 3-year.

A dependent variable in the model is the price-to-sales ratio as at the end of June 2011, denoted as P/S_i, and computed as a ratio of a stock price of i-th airline on June 30, 2011, to the annual total sales revenues of the same airline in the last full financial year ending before June 30, 2011 (extracted from its published consolidated income statement for the most recent annual period before June 30, 2011) divided by a total number of its common shares as at the end of June 2011.

The historical data on stock prices of individual airlines have been obtained from their periodic (annual and quarterly) financial reports, publicly available on their corporate websites. All the accounting data (necessary as inputs to both P/S multiples as well as financial statement ratios investigated in this chapter) have been extracted from the following corporate documents:

- "Aer Lingus Annual Report 2010."
- "Air Berlin Annual Report 2010."
- "AirFrance—KLM Annual Report 2010–11."
- "AMR Corporation 2010 Annual Report."
- "Cathay Pacific Airways Limited Annual Report 2010."
- "Delta Airlines Inc./DE/Form 10-K (Annual Report) Filed 02/16/11 for the Period Ending 12/31/10."
- "Lufthansa Annual Report 2010."
- "easyJet plc Annual Report and Accounts 2010."
- "Flybe Annual Report 2010/11."
- "British Airways Report and Accounts to December 2010."
- "Quantas Annual Report 2010."
- "Regional Express Holding Limited Annual Report for the Financial Year Ended 30 June 2010."
- "Ryanair Holdings plc Annual Report and Financial Statements 2011."
- "Spirit Airlines Inc. Annual Report Pursuant to Section 13 or 15(d) of the Securities Exchange Act of 1934 for the Fiscal Year ended December 31, 2011."
- "tigerairways.com Annual Report 2011. Financial Year Ended 31 March 2011."
- "United Continental Holdings Inc. Annual Report Pursuant to Section 13 or 15 (d) of the Securities Exchange Act of 1934 for the Fiscal Year ended December 31, 2010."

- "Virgin Blue Holdings Annual Report 2010."
- "WestJet Annual Report 2010."

Our purpose is to build a small-size model which captures the fundamental relationships between relative market values of individual airlines and their basic financial results. Thus, the preliminary set of candidate explanatory variables includes the following accounting ratios (computed on the ground of the annual reports published by individual airlines):

- *Margin*—cumulative consolidated operating profit in the last two full financial years ending before June 30, 2011, divided by cumulative consolidated sales revenues in the same 2-year period.
- *Leverage*—total consolidated liabilities and provisions at the end of the most recent financial year before June 30, 2011, divided by total consolidated assets at the same date.
- *Turnover*—consolidated sales revenues in the last full financial year ending before June 30, 2011, divided by total consolidated assets at the end of the same period.

The initial set of data for the model is shown in Table 9.1.

Table 9.1 The statistical data for all the observations of the dependent variable and the candidate explanatory variables in the model for P/S ratios[a] of selected airlines

Company	P/S_i[a]	Margin (%)	Leverage	Turnover
Aer Lingus Group plc	0,30	−3,3	0,56	0,67
Air Berlin plc	0,08	−1,5	0,79	1,57
Air France—KLM	0,10	−2,1	0,76	0,82
American Airlines Group Inc.	0,10	−1,1	1,16	0,88
Cathay Pacific Airways Limited	0,79	12,2	0,58	0,70
Delta Air Lines Inc.	0,21	−1,1	0,98	0,74
Deutsche Lufthansa	0,24	2,3	0,72	0,93
easyJet plc	0,17	3,4	0,63	0,74
Flybe Group plc	22,34	2,2	0,74	1,44
International Consolidated Airlines Group	30,72	−1,2	0,78	0,81
Quantas Airways Limited	0,30	0,8	0,70	0,69
Regional Express Holdings Limited	0,44	9,9	0,31	1,05
Ryanair Holdings plc	1,30	10,3	0,66	0,42
Spirit Airlines Inc.	0,45	10,5	1,22	1,64
Tiger Airways Holdings Limited	0,96	6,1	0,81	0,62
United Continental Holdings Inc.	0,26	−1,0	0,96	0,59
Virgin Australia Holdings Limited	0,21	−2,5	0,76	0,77
WestJet Airlines Ltd.	0,79	4,8	0,58	0,73

Source: Quarterly and annual reports of individual companies, author's computations
[a]Ratio of a stock price of i-th airline on June 30, 2011 to the annual total sales revenues of the same airline divided by a total number of its common shares as at the end of June 2011

9.3 Detection of Potential Outliers in the Dataset

The next step in our procedure of econometric modeling is to check for the presence of any outliers in our dataset. In order to do this, we will apply two complementary tools:

- Apply the "two-sigma range" to the individual variables.
- Apply the "two-sigma range" to the residuals obtained from the general regression.

9.3.1 The "Two-Sigma Range" Applied to the Individual Variables

Our first step in searching for outliers is to apply the "two-sigma range" to the individual variables. In order to reduce the risk of overlooking potential outliers, we set the upper and lower bounds based on their respective arithmetic means and two standard deviations. Such an approach results in narrower ranges whereby more potential outliers may be captured, at the small risk of misidentifying some true observations as outliers. Table 9.2 presents the results of the analysis.

Table 9.2 shows that in the case of three out of four investigated variables, the largest values in the sample exceed their respective upper bounds as determined by the "two-sigma range." However, in the case of *Leverage* and *Turnover*, the largest values in the sample exceed their respective thresholds only marginally. Thus, we apply our common sense here, and conclude that none of the observations of explanatory variables exceeds its respective upper threshold significantly. In contrast, in the case of the dependent variable, the maximum value in the sample exceeds its upper bound by a half, which means that it clearly deserves to be treated as the outlier.

As regards the lower bounds, the minimum observation of a variable is lower than its respective lower threshold only in the case of *Leverage*. However, similarly as for upper bounds, the distance between the lowest value in the sample and the

Table 9.2 "Two-sigma range" applied to the dependent variable and the candidate explanatory variables in the model for *P/S* ratios of selected airlines

Measure	P/S_i	Margin (%)	Leverage	Turnover
Arithmetic mean (AM)	3,32	2,7	0,76	0,88
Std. deviation (SD)	8,57	5,1	0,22	0,34
AM + 2 × SD	20,46	12,9	1,19	1,56
AM − 2 × SD	−13,82	−7,5	0,32	0,20
Max in the sample	30,72	12,2	1,22	1,64
Min in the sample	0,08	−3,3	0,31	0,42

Source: Quarterly and annual reports of individual companies, author's computations

lower bound is very small. Thus, we again apply our common sense here, and conclude that none of the observations of all four variables is lower than its respective lower threshold.

To sum up, so far we concluded that none of the observations of the explanatory variables is an outlier, while some of the observations of the dependent variable exceed the upper threshold. An inspection of data presented in Table 9.1 informs that apart from the maximum observation of the dependent variable (30,72), the second largest (22,34) also exceeds its upper bound. All the remaining observations of the dependent variable have values lower than 1,50 and thus lie strikingly far from both observations with the largest values. Accordingly, we conclude that these two largest observations of the dependent variable are outliers, and we remove them from the sample.

9.3.2 The "Two-Sigma Range" Applied to the Residuals from the General Regression

After removing two outlying observations (Flybe Group plc and International Consolidated Airlines Group) from the sample, we are left with sixteen remaining airlines. To check whether there are any significant multivariate outliers within this sample, we ran the general regression, in which all the candidate explanatory variables are present. Then we scrutinized the obtained residuals for the presence of any outliers.

After regressing P/S_i against all three candidate explanatory variables, we obtained the following model:

$$P/S_i = 0,62 + 5,19 \, Margin + 0,09 \, Leverage - 0,50 \, Turnover.$$

Then, for the obtained general regression, we computed the fitted values of the dependent variable and plotted them against the actual values of the dependent variable. This is shown in Fig. 9.1.

The visual inspection of Fig. 9.1 suggests that all observations lie near the core relationship. Also, as presented in Fig. 9.2, none of the residuals obtained from the general regression of P/S ratios seems to be outstanding. However, to formally confirm the lack of any multivariate outliers in the sample, we apply the "two-sigma range" to the obtained residuals. Similarly as in the case of individual variables, we set the upper and lower bounds on the ground of the respective arithmetic means and two (instead of three) standard deviations. Table 9.3 presents the results of the analysis.

The analysis presented in Table 9.3 corroborates that there are not any multivariate outliers in our dataset. Before we proceed further, our updated dataset looks as presented in Table 9.4.

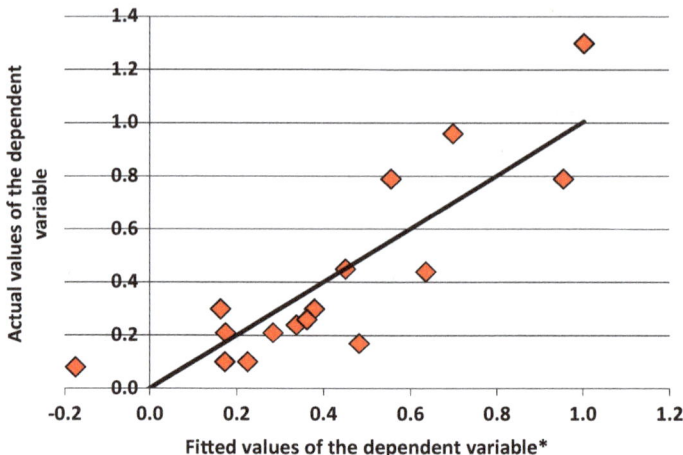

Fig. 9.1 Actual values of the dependent variable against the fitted values of the dependent variable from the general regression of *P/S* ratios of selected airlines. *From the general regression. Source: quarterly and annual reports of individual companies, author's computations

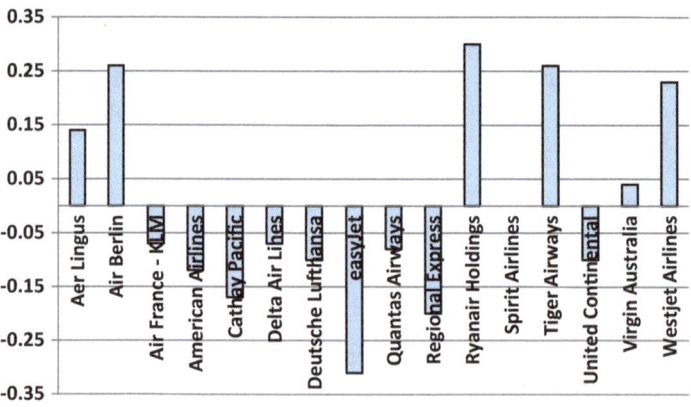

Fig. 9.2 The residuals obtained from the general regression of *P/S* ratios of selected airlines. Source: quarterly and annual reports of individual companies, author's computations

9.4   Selection of Explanatory Variables (from the Set of Candidates)

For our general model, which includes four variables presented in Table 9.5, we obtained the following results.

Given the results shown in the above table, we should remove *Leverage* from our set of explanatory variables, since the absolute value of its *t*-Statistic is below the

Table 9.3 "Two-sigma range" applied to the residuals obtained from the general regression of P/S ratios of selected airlines

Company	Residual from the general regression
Aer Lingus Group plc	0,14
Air Berlin plc	0,26
Air France—KLM	−0,07
American Airlines Group Inc.	−0,12
Cathay Pacific Airways Limited	−0,17
Delta Air Lines Inc.	−0,07
Deutsche Lufthansa	−0,10
easyJet plc	−0,31
Quantas Airways Limited	−0,08
Regional Express Holdings Limited	−0,20
Ryanair Holdings plc	0,30
Spirit Airlines Inc.	0,00
Tiger Airways Holdings Limited	0,26
United Continental Holdings Inc.	−0,10
Virgin Australia Holdings Limited	0,04
WestJet Airlines Ltd.	0,23
Arithmetic mean (AM)	0,00
Standard deviation (SD)	0,19
Upper bound (AM + 2 × SD)	0,37
Lower bound (AM − 2 × SD)	−0,37

Source: Quarterly and annual reports of individual companies, author's computations

Table 9.4 The final dataset for the general regression of P/S ratios of selected airlines

Company	P/S_i	Margin (%)	Leverage	Turnover
Aer Lingus Group plc	0,30	−3,3	0,56	0,67
Air Berlin plc	0,08	−1,5	0,79	1,57
Air France—KLM	0,10	−2,1	0,76	0,82
American Airlines Group Inc.	0,10	−1,1	1,16	0,88
Cathay Pacific Airways Limited	0,79	12,2	0,58	0,70
Delta Air Lines Inc.	0,21	−1,1	0,98	0,74
Deutsche Lufthansa	0,24	2,3	0,72	0,93
easyJet plc	0,17	3,4	0,63	0,74
Quantas Airways Limited	0,30	0,8	0,70	0,69
Regional Express Holdings Limited	0,44	9,9	0,31	1,05
Ryanair Holdings plc	1,30	10,3	0,66	0,42
Spirit Airlines Inc.	0,45	10,5	1,22	1,64
Tiger Airways Holdings Limited	0,96	6,1	0,81	0,62
United Continental Holdings Inc.	0,26	−1,0	0,96	0,59
Virgin Australia Holdings Limited	0,21	−2,5	0,76	0,77
WestJet Airlines Ltd.	0,79	4,8	0,58	0,73

Source: Quarterly and annual reports of individual companies, author's computations

critical value and is the smallest among all the explanatory variables. After removal of *Leverage*, we reestimate the parameters of the reduced model and repeat the procedure of the statistical significance analysis. The results of this re-estimation are presented below (Table 9.6).

We can see that both remaining explanatory variables are statistically significant at the 1% significance level, meaning that we completed the procedure of variable selection. We began with a general model which included three candidate explanatory variables and ended up with a specific model including two statistically significant explanatory variables.

Finally, the linear regression model of *P/S* ratios of selected airlines, obtained from the "general to specific modeling" procedure, looks as follows:

$$P/S_i = 0,68 + 5,09 \; Margin - 0,48 \; Turnover.$$

The model's *R*-squared equals 0,73 which means that it is capable of explaining a significant part of the intercompany differences in price-to-sales multiples, as at the end of June 2011. Such coefficient of determination suggests also that a model may

Table 9.5 Regression results for the general model

Variable	Coefficient	t-Statistic[b]
Intercept	0,62	2,972
Margin	5,19	4,977
Leverage	0,09	**0,358**
Turnover	−0,50	−2,903

R-squared: 0,73
Critical value of *t*-Statistic[a]: 3,055

Conclusion: remove *Leverage*

Source: Quarterly and annual reports of individual companies, author's computations
[a]For 16 observations, 3 explanatory variables and 1% significance level
[b]The smallest absolute value is noted

Table 9.6 Regression results for the reestimated model (i.e., after removing *Leverage* from the set of explanatory variables)

Variable	Coefficient	t-Statistic[b]
Intercept	0,68	4,728
Margin	5,09	5,251
Turnover	−0,48	**−3,073**

R-squared: 0,73
Critical value of *t*-Statistic[a]: 3,012

Conclusion: both explanatory variables are statistically significant at 1% significance level

Source: Quarterly and annual reports of individual companies, author's computations
[a]For 16 observations, 2 explanatory variables, and 1% significance level
[b]The smallest absolute value is noted

be useful in practice. However, before the model is applied, it must be verified for its statistical correctness. We will conduct such a verification in the following section.

9.5   Verification of the Obtained Model

We now apply the comprehensive procedure for verification of the resulting model, which will include the following tests:

- F-test for the general statistical significance of the whole model.
- Hellwig test for normality of distribution of residuals.
- Breusch–Pagan test for heteroscedasticity of residuals.
- t-Student test for symmetry of residuals.
- Maximum series length test for randomness of residuals.
- Ramsey's *RESET* test for the general specification of the model.
- Variance inflation factor test for multicollinearity of explanatory variables.

Since our model is a cross-sectional one, there is no need to test for the autocorrelation of residuals.

9.5.1   F-test for the General Statistical Significance of the Model

Our model was estimated on the basis of $n = 16$ observations. There are $k = 2$ explanatory variables and the coefficient of determination, $R^2 = 0,728$. We then calculate the empirical F statistic with the following results:

$$F = \frac{R^2}{1 - R^2} \frac{n - (k + 1)}{k} = \frac{0,728}{1 - 0,728} \frac{16 - (2 + 1)}{2} = 17,40.$$

For $n = 16$ and $k = 2$, we obtain:

$$m_1 = k = 2,$$
$$m_2 = n - (k + 1) = 16 - (2 + 1) = 13.$$

For the level of statistical significance set at $\alpha = 0,05$, and computed values of m_1 and m_2, we obtain the critical value of F^* from the Fisher–Snedecor table or from the Excel function $=$ F.INV.RT(0.05,2,13), which equals 3,81.

We then compare the value of empirical statistic F with its critical value F^*. In our case $F > F^*$, which means that at least one slope parameter in our model is statistically different than zero.

9.5.2   Hellwig Test for Normality of Distribution of Residuals

To check the normality of residuals' distribution in our model by means of the Hellwig test, the first thing is to compute those residuals. This is presented in Table 9.7.

According to the Hellwig test procedure, we need to sort all the residuals in order of their increasing values and then standardize the residuals on the basis of their arithmetic average and their standard deviation. The next step is to obtain the normal distribution frequency from the table of normal distribution (or from Excel). We show the results of these three steps in Table 9.8.

In the next step, we create $n = 16$ cells, by dividing the range [0, 1] into n equal sub-ranges. For $n = 16$, the individual cell width is $0,063 = 1/16$. Then, on the basis of the normal distribution frequencies appointed for the standardized residuals, we compute the number of residuals falling into each individual cell. Finally, we compute the value of k_e, which is the number of empty cells. All these three steps are presented in Table 9.9.

As the data in Table 9.9 show, the number of empty cells $k_e = 7$. Then we obtain the critical Hellwig test values of k_1 and k_2, for $n = 16$ observations and for the assumed level of statistical significance α (we assume $\propto = 0,05$). In the appropriate table (found in the Appendix A4), we find that $k_1 = 3$ and $k_2 = 8$.

Table 9.7   Computation of residuals from the model of P/S ratios of selected airlines

Company	P/S_i (actual values)	$\widehat{P/S_i}$ (fitted values)[a]	$e_i = P/S_i - \widehat{P/S_i}$
Aer Lingus Group plc	0,30	0,16	0,14
Air Berlin plc	0,08	−0,18	0,26
Air France—KLM	0,10	0,17	−0,07
American Airlines Group Inc.	0,10	0,22	−0,12
Cathay Pacific Airways Limited	0,79	0,96	−0,17
Delta Air Lines Inc.	0,21	0,28	−0,07
Deutsche Lufthansa	0,24	0,34	−0,10
easyJet plc	0,17	0,48	−0,31
Quantas Airways Limited	0,30	0,38	−0,08
Regional Express Holdings Limited	0,44	0,64	−0,20
Ryanair Holdings plc	1,30	1,00	0,30
Spirit Airlines Inc.	0,45	0,45	0,00
Tiger Airways Holdings Limited	0,96	0,70	0,26
United Continental Holdings Inc.	0,26	0,36	−0,10
Virgin Australia Holdings Limited	0,21	0,17	0,04
WestJet Airlines Ltd.	0,79	0,56	0,23

Source: Quarterly and annual reports of individual companies, author's computations
[a]Computed according to the obtained model, $P/S_i = 0,68 + 5,09 \, Margin - 0,48 \, Turnover$

Table 9.8 The results of the residuals sorting, standardization, and derivation of the normal distribution frequencies

Residuals $(e_i = P/S_i - \widehat{P/S_i})$	Sorted residuals e_{is}	Standardized residuals e'_{is}	Normal distribution frequency $F(e_{is})$[a]
0,14	−0,32	−1,735	0,041
0,26	−0,23	−1,247	0,106
−0,07	−0,17	−0,922	0,178
−0,12	−0,10	−0,542	0,294
−0,17	−0,09	−0,488	0,313
−0,07	−0,08	−0,434	0,332
−0,10	−0,08	−0,434	0,332
−0,31	−0,08	−0,434	0,332
−0,08	−0,06	−0,325	0,372
−0,20	0,03	0,163	0,565
0,30	0,03	0,163	0,565
0,00	0,12	0,651	0,742
0,26	0,22	1,193	0,884
−0,10	0,24	1,301	0,903
0,04	0,27	1,464	0,928
0,23	0,30	1,627	0,948
Arithmetic mean ($\bar{e}$)	0	0	–
Std. deviation (S)	0,18	1	–

Source: Quarterly and annual reports of individual companies, author's computations
[a]The values derived from the table for normal distribution frequency or from the spreadsheet

Since $k_1 \leq k_e \leq k_2$, we conclude that no statistical evidence was found against the normality of the residuals from our model of P/S ratios of selected airlines.

9.5.3 Breusch–Pagan Test for Heteroscedasticity of Residuals

In model estimated for Nokia's quarterly sales revenues, presented in the preceding chapter, we tested for heteroscedasticity of residuals by means of ARCH-LM test. However, ARCH-LM test is not useful here because it is applicable only for time-series models. Instead, we should apply the alternative tool, which is Breusch–Pagan test.

The first step to do is to prepare the data by computing the values of r_n. These computations are shown in Table 9.10. Then we need to regress r_n against the set of explanatory variables from our model. The results of this regression are presented in Table 9.11.

The regression of r_n against *Margin* and *Turnover*, presented in Table 9.11, was estimated on the basis of 16 observations. This means that for Breusch–Pagan test of our model $n = 16$, there are two explanatory variables (which means that $k = 2$)

Table 9.9   The computation of the number of empty cells for Hellwig test

Cell	Number of residuals falling into the cell[a]
<0,000; 0,063)	1
<0,063; 0,125)	1
<0,125; 0,188)	1
<0,188; 0,250)	0
<0,250; 0,313)	1
<0,313; 0,375)	5
<0,375; 0,438)	0
<0,438; 0,500)	0
<0,500; 0,563)	0
<0,563; 0,625)	2
<0,625; 0,688)	0
<0,688; 0,750)	1
<0,750; 0,813)	0
<0,813; 0,875)	0
<0,875; 0,938)	3
<0,938; 1,000>	1
The number of empty cells (k_e)	7

Source: Quarterly and annual reports of individual companies, author's computations
[a]On the basis of the data from the last column of Table 9.8

and the coefficient of determination $R^2 = 0,225$. Putting these numbers into the formula for the empirical statistic F produces the following results:

$$F = \frac{R^2}{1-R^2} \frac{n-(k+1)}{k} = \frac{0,225}{1-0,225} \frac{16-(2+1)}{2} = 1,88.$$

Now we need to derive the critical value of F^*. For $n = 16$ and $k = 2$, we obtain:

$$m_1 = k = 2,$$
$$m_2 = n - (k+1) = 16 - (2+1) = 13.$$

For the level of statistical significance set at $\alpha = 0,05$, and computed values of m_1 and m_2, we obtain the critical value of F^* from the Fisher–Snedecor table or from the Excel function = F.INV.RT(0.05,2,13), which equals 3,81.

Finally, we need to compare the value of the empirical statistic F with its critical value F^*. In our case $F < F^*$, which means that H_0 is not rejected, and we conclude that there is no significant heteroscedasticity of the residuals in our model.

Table 9.10 The computations of r_n for Breusch–Pagan test (for heteroscedasticity of residuals)

Residuals e_i	Squared residuals e_i^2	$r_n = \frac{e_i^2}{S^2} - 1$
0,12	0,014	−0,588
0,24	0,058	0,706
−0,08	0,006	−0,824
−0,09	0,008	−0,765
−0,17	0,029	−0,147
−0,06	0,004	−0,882
−0,10	0,010	−0,706
−0,32	0,102	2,000
−0,08	0,006	−0,824
−0,23	0,053	0,559
0,30	0,090	1,647
0,03	0,001	−0,971
0,27	0,073	1,147
−0,08	0,006	−0,824
0,03	0,001	−0,971
0,22	0,048	0,412

Variance of residuals: $S^2 = 0,034$

Source: Quarterly and annual reports of individual companies, author's computations

Table 9.11 Results of regressing r_n against the explanatory variables from the tested model

Variable	Coefficient	t-Statistic
Intercept	0,18	0,266
Margin	8,43	1,829
Turnover	−0,59	−0,784

R-squared: 0,225

Source: Quarterly and annual reports of individual companies, author's computations

9.5.4 t-Student Test for Symmetry of Residuals

In checking the symmetry of residuals in our model, the first thing to do is to compute the number of positive residuals (denoted as m_{e1}) and the number of negative residuals (denoted as m_{e2}). This is shown in Table 9.12.

Then we calculate the empirical statistics as follows:

$$t_{e1} = \frac{\left|\frac{m_{e1}}{n} - \frac{1}{2}\right|}{\sqrt{\frac{\frac{m_{e1}}{n}\left(1 - \frac{m_{e1}}{n}\right)}{n-1}}} = \frac{\left|\frac{7}{16} - \frac{1}{2}\right|}{\sqrt{\frac{\frac{7}{16}\left(1 - \frac{7}{16}\right)}{16-1}}} = 0,488,$$

Table 9.12 The computations for the symmetry test (for symmetry of residuals from the model of P/S ratios of selected airlines)

Residuals e_i	Sign of residual (1 – positive, 0 – otherwise)	Sign of residual (1 – negative, 0 – otherwise)
0,12	1	0
0,24	1	0
−0,08	0	1
−0,09	0	1
−0,17	0	1
−0,06	0	1
−0,10	0	1
−0,32	0	1
−0,08	0	1
−0,23	0	1
0,30	1	0
0,03	1	0
0,27	1	0
−0,08	0	1
0,03	1	0
0,22	1	0
$m_{e1}=$	7	–
$m_{e2}=$	–	9

Source: Quarterly and annual reports of individual companies, author's computations

$$t_{e2} = \frac{\left|\frac{m_{e1}}{n} - \frac{1}{2}\right|}{\sqrt{\frac{\frac{m_{e1}}{n}\left(1-\frac{m_{e1}}{n}\right)}{n-1}}} = \frac{\left|\frac{9}{16} - \frac{1}{2}\right|}{\sqrt{\frac{\frac{9}{16}\left(1-\frac{9}{16}\right)}{16-1}}} = 0,488.$$

The next step is to obtain the critical value of t_t, from the t-Student distribution table or from Excel for $n - 1 = 15$ degrees of freedom and for the assumed level of statistical significance $\alpha = 5\%$ (=T.INV(0.05,15)). This value of t_t equals 2,131. Given that in our model $t_{e1} < t_t$ as well as $t_{e2} < t_t$, we may conclude that no statistical evidence was found against the symmetry of the residuals from our model of P/S ratios of selected airlines.

9.5.5   Maximum Series Length Test for Randomness of Residuals

In the model estimated for Nokia's quarterly sales revenues, presented in the preceding chapter, we tested for randomness of residuals on the basis of the actual order of residuals. This was so because in time-series models, the non-randomness of residuals, if it exists, tends to be correlated with time (e.g., in time-series models the series of positive residuals clustered around "older" observations may be followed by the series of negative residuals related to "newer" observations). In contrast, in cross-sectional models, the individual observations do not have any order (i.e., they are not related to each other in the same way as the time-series data). However, they may still present some non-randomness, which may be correlated with one or more of the variables in the model (e.g., positive/negative residuals may be clustered around observations with low/high values of dependent variable). Thus, it is still advisable to check for the randomness of residuals in cross-sectional models. However, the test must be preceded by sorting the residuals in order of increasing or decreasing values of model variables.

For the maximum series length test, applied to the residuals from our model of P/S ratios of selected airlines, we assume that the pattern of signs of residuals (if any) is related to the dependent variable. Thus, before running the test, we sort all the residuals in order of increasing values of the dependent variable. This sorting is presented in Table 9.13.

After sorting the observations, we denote the positive residuals as "A" and negative residuals as "B." Then we compute the length of the longest series of residuals of the same sign (k_e). This is shown in the following table (Table 9.14).

We then obtain the critical value of k_t from the maximum series length table (included in the Appendix A6). For 16 residuals ($n = 24$) and for the assumed level of statistical significance α of 5%, we find that $k_t = 6$. Given that in our model $k_e < k_t$, we may conclude that no statistical evidence was found against the randomness of the residuals form our model of P/S ratios of selected airlines.

Table 9.13 Sorting of residuals (from the model of P/S ratios of selected airlines) in order of increasing values of the dependent variable

Original order of observations		Observations sorted in order of increasing values of dependent variable	
Dependent variable P/S_i	Residuals e_i	Dependent variable P/S_i	Residuals e_i
0,30	0,12	0,08	0,24
0,08	0,24	0,10	−0,08
0,10	−0,08	0,10	−0,09
0,10	−0,09	0,17	−0,32
0,79	−0,17	0,21	−0,06
0,21	−0,06	0,21	0,03
0,24	−0,10	0,24	−0,10
0,17	−0,32	0,26	−0,08
0,30	−0,08	0,30	0,12
0,44	−0,23	0,30	−0,08
1,30	0,30	0,44	−0,23
0,45	0,03	0,45	0,03
0,96	0,27	0,79	−0,17
0,26	−0,08	0,79	0,22
0,21	0,03	0,96	0,27
0,79	0,22	1,30	0,30

Source: Quarterly and annual reports of individual companies, author's computations

Table 9.14 The computations for the maximum series length test (for randomness of residuals from our model of P/S ratios of selected airlines)

Residuals e_i	Sign of residual ("A" – positive, "B" – negative)	Length of the series of residuals of the same sign
0,24	A	1
−0,08	B	
−0,09	B	
−0,32	B	
−0,06	B	4
0,03	A	1
−0,10	B	
−0,08	B	2
0,12	A	1
−0,08	B	
−0,23	B	2
0,03	A	1
−0,17	B	1
0,22	A	
0,27	A	
0,30	A	3
Length of the longest series of residuals (k_e)		4

Source: Quarterly and annual reports of individual companies, author's computations

9.5.6 Ramsey's RESET Test for the General Specification of the Model

In applying the Ramsey's *RESET* test, the first thing to do is to prepare the data by computing the fitted values of the dependent variable as well as the values of auxiliary explanatory variables. We will add two auxiliary variables:

- Squared fitted values of the dependent variable obtained from our model (denoted as $FITTED^2$).
- Cubic fitted values of the dependent variable obtained from our model (denoted as $FITTED^3$).

These data are presented in Table 9.15. Table 9.16, in turn, presents the results of auxiliary regression estimated for Ramsey's RESET test for our model of P/S ratios of selected airlines.

We know from the F-test for the general significance that the R-squared of our model equals 0,728. Table 9.16, in turn, informs us that the R-squared of the auxiliary regression equals 0,788. Our tested model was estimated on the basis of 16 observations ($n = 16$), and it includes two explanatory variables (which means that $k = 2$). Therefore, our empirical F statistic is computed as follows:

$$F = \frac{\left(R_2^2 - R_1^2\right)/2}{\left(1 - R_2^2\right)/(n - (k+3))} = \frac{(0,788 - 0,728)/2}{(1 - 0,788)/(16 - (2+3))} = 1,56.$$

Table 9.15 The data for Ramsey's *RESET* test for correctness of specification of *P/S* ratios of selected airlines

Company	Fitted P/S_i	$FITTED^2$	$FITTED^3$
Aer Lingus Group plc	0,18	0,0337	0,0062
Air Berlin plc	−0,16	0,0249	−0,0039
Air France—KLM	0,18	0,0310	0,0055
American Airlines Group Inc.	0,19	0,0375	0,0073
Cathay Pacific Airways Limited	0,96	0,9211	0,8840
Delta Air Lines Inc.	0,27	0,0705	0,0187
Deutsche Lufthansa	0,34	0,1181	0,0406
easyJet plc	0,49	0,2408	0,1182
Quantas Airways Limited	0,38	0,1467	0,0562
Regional Express Holdings Limited	0,67	0,4535	0,3054
Ryanair Holdings plc	1,00	0,9932	0,9898
Spirit Airlines Inc.	0,42	0,1752	0,0733
Tiger Airways Holdings Limited	0,69	0,4712	0,3235
United Continental Holdings Inc.	0,34	0,1170	0,0400
Virgin Australia Holdings Limited	0,18	0,0315	0,0056
WestJet Airlines Ltd.	0,57	0,3218	0,1825

Source: Quarterly and annual reports of individual companies, author's computations

Table 9.16 Auxiliary regression estimated for Ramsey's *RESET* test conducted for our model of *P/S* ratios of selected airlines

Variable	Coefficient	*t*-Statistic
Intercept	0,16	0,427
Margin	−0,39	−0,106
Turnover	−0,09	−0,309
$FITTED^2$	2,10	1,018
$FITTED^3$	−1,07	−0,688
R-squared: 0,788		

Source: Quarterly and annual reports of individual companies, author's computations

Now we need to obtain the critical value of F^*. For $n = 16$ and $k = 2$, we obtain:

$$m_1 = 2,$$
$$m_2 = n - (k + 3) = 16 - (2 + 3) = 11.$$

For the level of statistical significance set at $\alpha = 0,05$, and determined values of m_1 and m_2, we obtain the critical value of F^* from the Fisher–Snedecor table or from the Excel function = F.INV.RT(0.05,2,11), which equals 3,98.

Finally, we compare the value of the empirical statistic F with its critical value F^*. In our case $F < F^*$, which means that none of the two auxiliary variables is statistically significant. Therefore, we can conclude that our model of *P/S* ratios of selected airlines is correctly specified.

Table 9.17 Results of variance inflation factor test run for explanatory variables in our model of *P/S* ratios of selected airlines

Auxiliary regression	*R*-squared	*VIF_i* statistic
Margin against *Turnover*	0,006	1,01
Critical value of *VIF_i* statistic (as a "rule of thumb"): 5,00		

Source: Quarterly and annual reports of individual companies, author's computations

9.5.7   Variance Inflation Factor Test for the Multicollinearity of Explanatory Variables

Finally, we test the potential multicollinearity of explanatory variables in our model. It includes only two explanatory variables, which means that we need to run only one test for multicollinearity (i.e., regress one of the explanatory variables against the another one). The results of the auxiliary regression are presented in Table 9.17. As we can see, our model is free from multicollinearity of explanatory variables, because the empirical *VIF_i* statistic has value below the critical threshold.

9.6   Evaluation of Model Usefulness in Identifying Overvalued and Undervalued Stocks

In this section, we evaluate a model's capability of picking overvalued and undervalued airline companies. In the analysis, we take the following assumptions:

- Any airline whose actual *P/S* ratio at the end of June 2011 exceeds its fundamental (fitted) ratio (which implies a positive residual from our regression) is considered to be overvalued, i.e., priced by the stock market at a value which exceeds its true value.
- Any airline whose actual *P/S* ratio at the end of June 2011 lies below its fundamental (fitted) ratio (which implies a negative residual from our regression) is considered to be undervalued, i.e., priced by the stock market at a value which understates its true value.

To evaluate the model usefulness in building investment portfolios, we compare stock returns of two alternative portfolios:

- Portfolio of overvalued stocks, which includes all the airlines with positive regression residuals.
- Portfolio of undervalued stocks, which includes all the airlines with negative regression residuals.

We expect overvalued and undervalued stocks to generate below-average and above-average stock returns, respectively. The comparative analysis assumes three alternative investment horizons:

- One-year horizon, which assumes purchase of stocks at the end of June 2011 and sale of stocks at the end of June 2012.

- Two-year horizon, which assumes purchase of stocks at the end of June 2011 and sale of stocks at the end of June 2013,
- Two-year horizon, which assumes purchase of stocks at the end of June 2011 and sale of stocks at the end of June 2014.

Table 9.18 compares the stock returns of individual airlines as well as our two alternative portfolios.

Table 9.18 Stock returns of individual airlines as well as two alternative investment portfolios formed on the basis of the estimated model

	Companies[a]	Residuals from our model	Stock price changes		
			One year (June 2011 to June 2012) (%)	Two year (June 2011 to June 2013)	Three year (June 2011 to June 2014)
Portfolio of overvalued stocks	Ryanair Holdings plc	0,30	25,2	111,4	115,8
	Tiger Airways Holdings Limited	0,27	−36,4	−45,5	−60,9
	Air Berlin plc	0,24	−49,4	−47,2	−63,9
	WestJet Airlines Ltd.	0,22	15,5	53,0	95,5
	Aer Lingus Group plc	0,12	56,5	142,8	92,0
	Virgin Australia Holdings Limited	0,03	37,9	55,2	41,4
	Spirit Airlines Inc.	0,03	64,2	152,3	399,4
Portfolio of undervalued stocks	Delta Air Lines Inc.	−0,06	22,3	169,1	374,8
	Air France—KLM	−0,08	−48,6	−27,8	−4,2
	United Continental Holdings Inc.	−0,08	4,2	92,3	156,0
	Quantas Airways Limited	−0,08	−38,4	−31,9	−27,6
	American Airlines Group Inc.	−0,09	83,7	210,1	522,6
	Deutsche Lufthansa	−0,10	−27,0	7,1	−5,6
	Cathay Pacific Airways Limited	−0,17	−28,9	−20,5	−18,5
	Regional Express Holdings Limited	−0,23	30,3	32,6	0,0
	easyJet plc	−0,32	52,9	284,7	252,9
Returns of the whole portfolio of overvalued stocks[b]			16,2	60,3	88,5
Returns of the whole portfolio of undervalued stocks[c]			5,6	79,5	138,9

Source: Quarterly and annual reports of individual companies, author's computations
[a]Sorted in order of decreasing regression residuals
[b]Arithmetic averages of the returns of individual companies included in the portfolio of overvalued stocks (i.e., companies with positive regression residuals)
[c]Arithmetic averages of the returns of individual companies included in the portfolio of undervalued stocks (i.e., companies with negative regression residuals)

As might be seen, the obtained model seems not to be very useful in screening stocks for investments with relatively short-term horizons. The 1 year nominal return of overvalued stocks (16,2%) significantly exceeds the return earned by the alternative portfolio consisting of undervalued companies (5,6%). This would not be surprising for a majority of stock investment experts, who know that in the short-run stock price changes are formed by so many variables (including psychological and other qualitative factors) that they cannot be captured by such simple models. However, the results obtained for longer investment horizons (2-year and 3-year) are much more promising. Here the portfolio of undervalued stocks "beats" the alternative one by almost 20% points in a 2-year timeframe and as many as 50% points in the 3-year horizon. Thus, it seems that the proposed econometric model (when updated with new data) may constitute an informative tool, useful in building long-term stock investment portfolios.

Appendix: Statistical Tables

A1. Critical Values for *F*-statistic for $\alpha = 0, 05$

m_2	m_1								
	1	2	3	4	5	6	7	8	9
1	161	200	216	225	230	234	237	239	241
2	18,5	19,0	19,2	19,2	19,3	19,4	19,4	19,4	19,4
3	10,1	9,55	9,28	9,12	9,01	8,94	8,89	8,85	8,81
4	7,71	6,94	6,59	6,39	6,26	6,16	6,09	6,04	6,00
5	6,61	5,79	5,41	5,19	5,05	4,95	4,88	4,82	4,77
6	5,99	5,14	4,76	4,53	4,39	4,28	4,21	4,15	4,10
7	5,59	4,74	4,35	4,12	3,97	3,87	3,79	3,73	3,68
8	5,32	4,46	4,07	3,84	3,69	3,58	3,50	3,44	3,39
9	5,12	4,26	3,86	3,63	3,48	3,37	3,29	3,23	3,18
10	4,96	4,10	3,71	3,48	3,33	3,22	3,14	3,07	3,02
11	4,84	3,98	3,59	3,36	3,20	3,09	3,01	2,95	2,90
12	4,75	3,89	3,49	3,26	3,11	3,00	2,91	2,85	2,80
13	4,67	3,81	3,41	3,18	3,03	2,92	2,83	2,77	2,71
14	4,60	3,74	3,34	3,11	2,96	2,85	2,76	2,70	2,65
15	4,54	3,68	3,29	3,06	2,90	2,79	2,71	2,64	2,59
16	4,49	3,63	3,24	3,01	2,85	2,74	2,66	2,59	2,54
17	4,45	3,59	3,20	2,96	2,81	2,70	2,61	2,55	2,49
18	4,41	3,55	3,16	2,93	2,77	2,66	2,58	2,51	2,46
19	4,38	3,52	3,13	2,90	2,74	2,63	2,54	2,48	2,42
20	4,35	3,49	3,10	2,87	2,71	2,60	2,51	2,45	2,39
21	4,32	3,47	3,07	2,84	2,68	2,57	2,49	2,42	2,37
22	4,30	3,44	3,05	2,82	2,66	2,55	2,46	2,40	2,34
23	4,28	3,42	3,03	2,80	2,64	2,53	2,44	2,37	2,32
24	4,26	3,40	3,01	2,78	2,62	2,51	2,42	2,36	2,30
25	4,24	3,39	2,99	2,76	2,60	2,49	2,40	2,34	2,28

(continued)

© Springer International Publishing AG 2018
J. Welc, P.J.R. Esquerdo, *Applied Regression Analysis for Business*,
https://doi.org/10.1007/978-3-319-71156-0

m_2	m_1								
	1	2	3	4	5	6	7	8	9
26	4,23	3,37	2,98	2,74	2,59	2,47	2,39	2,32	2,27
27	4,21	3,35	2,96	2,73	2,57	2,46	2,37	2,31	2,25
28	4,20	3,34	2,95	2,71	2,56	2,45	2,36	2,29	2,24
29	4,18	3,33	2,93	2,70	2,55	2,43	2,35	2,28	2,22
30	4,17	3,32	2,92	2,69	2,53	2,42	2,33	2,27	2,21
40	4,08	3,23	2,84	2,61	2,45	2,34	2,25	2,18	2,12
60	4,00	3,15	2,76	2,53	2,37	2,25	2,17	2,10	2,04
100	3,94	3,09	2,70	2,46	2,31	2,19	2,10	2,03	1,97
200	3,89	3,04	2,65	2,42	2,26	2,14	2,06	1,98	1,93
1.000	3,85	3,00	2,61	2,38	2,22	2,11	2,02	1,95	1,89

m_2	m_1								
	10	12	14	15	20	30	60	120	∞
1	242	244	245	246	248	250	252	253	254
2	19,4	19,4	19,4	19,4	19,4	19,5	19,5	19,5	19,5
3	8,79	8,74	8,71	8,70	8,66	8,62	8,57	8,55	8,53
4	5,96	5,91	5,87	5,86	5,80	5,75	5,69	5,66	5,63
5	4,74	4,68	4,64	4,62	4,56	4,50	4,43	4,40	4,36
6	4,06	4,00	3,96	3,94	3,87	3,81	3,74	3,70	3,67
7	3,64	3,57	3,53	3,51	3,44	3,38	3,30	3,27	3,23
8	3,35	3,28	3,24	3,22	3,15	3,08	3,01	2,97	2,93
9	3,14	3,07	3,03	3,01	2,94	2,86	2,79	2,75	2,71
10	2,98	2,91	2,86	2,84	2,77	2,70	2,62	2,58	2,54
11	2,85	2,79	2,74	2,72	2,65	2,57	2,49	2,45	2,40
12	2,75	2,69	2,64	2,62	2,54	2,47	2,38	2,34	2,30
13	2,67	2,60	2,55	2,53	2,46	2,38	2,30	2,25	2,21
14	2,60	2,53	2,48	2,46	2,39	2,31	2,22	2,18	2,13
15	2,54	2,48	2,42	2,40	2,33	2,25	2,16	2,11	2,07
16	2,49	2,42	2,37	2,35	2,28	2,19	2,11	2,06	2,01
17	2,45	2,38	2,33	2,31	2,23	2,15	2,06	2,01	1,96
18	2,41	2,34	2,29	2,27	2,19	2,11	2,02	1,97	1,92
19	2,38	2,31	2,26	2,23	2,16	2,07	1,98	1,93	1,88
20	2,35	2,28	2,22	2,20	2,12	2,04	1,95	1,90	1,84
21	2,32	2,25	2,20	2,18	2,10	2,01	1,92	1,87	1,81
22	2,30	2,23	2,17	2,15	2,07	1,98	1,89	1,84	1,78
23	2,27	2,20	2,15	2,13	2,05	1,96	1,86	1,81	1,76
24	2,25	2,18	2,13	2,11	2,03	1,94	1,84	1,79	1,73
25	2,24	2,16	2,11	2,09	2,01	1,92	1,82	1,77	1,71
26	2,22	2,15	2,09	2,07	1,99	1,90	1,80	1,75	1,69
27	2,20	2,13	2,08	2,06	1,97	1,88	1,79	1,73	1,67
28	2,19	2,12	2,06	2,04	1,96	1,87	1,77	1,71	1,65
29	2,18	2,10	2,05	2,03	1,94	1,85	1,75	1,70	1,64

(continued)

m_2	m_1								
	10	12	14	15	20	30	60	120	∞
30	2,16	2,09	2,04	2,01	1,93	1,84	1,74	1,68	1,62
40	2,08	2,00	1,95	1,92	1,84	1,74	1,64	1,58	1,51
60	1,99	1,92	1,86	1,84	1,75	1,65	1,53	1,47	1,39
100	1,93	1,85	1,79	1,77	1,68	1,57	1,45	1,38	1,28
200	1,88	1,80	1,74	1,72	1,62	1,52	1,39	1,30	1,19
1.000	1,84	1,76	1,70	1,68	1,58	1,47	1,33	1,24	1,08

A2. Critical Values for t-statistic

Degrees of freedom	α							
	0,80	0,60	0,40	0,20	0,10	0,05	0,02	0,01
1	0,325	0,727	1,376	3,078	6,314	12,706	31,821	63,657
2	0,289	0,617	1,061	1,886	2,920	4,303	6,965	9,925
3	0,277	0,584	0,978	1,638	2,353	3,182	5,541	5,841
4	0,271	0,569	0,941	1,533	2,132	2,776	3,747	4,604
5	0,267	0,559	0,920	1,476	2,015	2,571	3,365	4,032
6	0,265	0,553	0,906	1,440	1,943	2,447	3,143	3,707
7	0,263	0,549	0,896	1,415	1,895	2,365	2,998	3,499
8	0,262	0,546	0,889	1,397	1,860	2,306	2,896	3,355
9	0,261	0,543	0,883	1,383	1,833	2,262	2,821	3,250
10	0,260	0,542	0,879	1,372	1,812	2,228	2,764	3,169
11	0,260	0,540	0,876	1,363	1,796	2,201	2,718	3,106
12	0,259	0,539	0,873	1,356	1,782	2,179	2,681	3,055
13	0,259	0,538	0,870	1,350	1,771	2,160	2,650	3,012
14	0,258	0,537	0,868	1,345	1,760	2,145	2,624	2,977
15	0,258	0,536	0,866	1,341	1,753	2,131	2,602	2,947
16	0,258	0,535	0,865	1,337	1,746	2,120	2,583	2,921
17	0,257	0,534	0,863	1,333	1,740	2,110	2,565	2,898
18	0,257	0,534	0,862	1,330	1,734	2,101	2,552	2,878
19	0,257	0,533	0,861	1,328	1,729	2,093	2,539	2,861
20	0,257	0,533	0,860	1,325	1,725	2,086	2,528	2,845
21	0,257	0,532	0,859	1,323	1,721	2,080	2,518	2,831
22	0,256	0,532	0,858	1,321	1,717	2,074	2,508	2,819
23	0,256	0,532	0,858	1,319	1,714	2,069	2,500	2,807
24	0,256	0,531	0,857	1,318	1,711	2,064	2,492	2,797
25	0,256	0,531	0,856	1,316	1,708	2,060	2,485	2,787
26	0,256	0,531	0,856	1,315	1,706	2,056	2,479	2,779
27	0,256	0,531	0,855	1,314	1,703	2,052	2,473	2,771
28	0,256	0,530	0,855	1,313	1,701	2,048	2,467	2,763

(continued)

Degrees of freedom	α							
	0,80	0,60	0,40	0,20	0,10	0,05	0,02	0,01
29	0,256	0,530	0,854	1,311	1,699	2,045	2,462	2,756
30	0,256	0,530	0,854	1,310	1,697	2,042	2,457	2,750
40	0,255	0,529	0,851	1,303	1,684	2,021	2,423	2,704
60	0,254	0,527	0,848	1,296	1,671	2,000	2,390	2,660
120	0,254	0,526	0,845	1,289	1,658	1,980	2,358	2,617
∞	0,253	0,524	0,842	1,282	1,645	1,960	2,326	2,576

A3. Critical Values for Chi-squared Statistic

Degrees of freedom	α					
	0,1	0,05	0,025	0,01	0,005	0,001
1	2,70	3,84	5,02	6,63	7,87	10,83
2	4,60	5,99	7,37	9,21	10,59	13,82
3	6,25	7,81	9,34	11,34	12,83	16,27
4	7,77	9,48	11,14	13,27	14,86	18,47
5	9,23	11,07	12,83	15,08	16,75	20,52
6	10,64	12,59	14,44	16,81	18,54	22,46
7	12,01	14,06	16,01	18,47	20,27	24,32
8	13,36	15,50	17,53	20,09	21,95	26,12
9	14,68	16,91	19,02	21,66	23,58	27,88
10	15,98	18,30	20,48	23,20	25,18	29,59
11	17,27	19,67	21,92	24,72	26,75	31,26
12	18,54	21,02	23,33	26,21	28,30	32,91
13	19,81	22,36	24,73	27,68	29,82	34,53
14	21,06	23,68	26,11	29,14	31,31	36,12
15	22,30	24,99	27,48	30,57	32,80	37,70
16	23,54	26,29	28,84	32,00	34,26	39,25
17	24,76	27,58	30,19	33,40	35,71	40,79
18	25,98	28,86	31,52	34,80	37,15	42,31
19	27,20	30,14	32,85	36,19	38,58	43,82
20	28,41	31,41	34,17	37,56	39,99	45,31
21	29,61	32,67	35,47	38,93	41,40	46,80
22	30,81	33,92	36,78	40,28	42,79	48,27
23	32,00	35,17	38,07	41,63	44,18	49,73
24	33,19	36,41	39,36	42,98	45,55	51,18
25	34,38	37,65	40,64	44,31	46,92	52,62
26	35,56	38,88	41,92	45,64	48,29	54,05
27	36,74	40,11	43,19	46,96	49,64	55,48
28	37,91	41,33	44,46	48,27	50,99	56,89

(continued)

Degrees of freedom	α					
	0,1	0,05	0,025	0,01	0,005	0,001
29	39,08	42,55	45,72	49,58	52,33	58,30
30	40,26	43,77	46,98	50,89	53,67	59,70
40	51,81	55,76	59,34	63,69	66,77	73,40
50	63,17	67,50	71,42	76,15	79,49	86,66
60	74,40	79,08	83,30	88,38	91,95	99,61
70	85,53	90,53	95,02	100,43	104,21	112,32
80	96,58	101,88	106,63	112,33	116,32	124,84
90	107,56	113,15	118,14	124,12	128,30	137,21
100	118,50	124,34	129,56	135,81	140,17	149,45

A4. Critical Values for Hellwig Test

n	α							
	0,10		0,05		0,01		0,005	
7	0	3	0	4	0	4	0	4
8	1	4	0	4	0	5	0	5
9	1	4	1	5	0	5	0	5
10	1	5	1	5	0	6	0	6
11	2	5	1	6	1	6	0	6
12	2	6	2	6	1	7	1	7
13	2	6	2	6	1	7	1	7
14	2	6	2	7	1	8	1	8
15	3	7	2	7	2	8	1	8
16	3	7	3	8	2	9	2	9
17	3	8	3	8	2	9	2	9
18	4	8	3	9	2	9	2	10
19	4	9	4	9	3	10	2	10
20	4	9	4	9	3	10	3	11
21	5	9	4	10	3	11	3	11
22	5	10	5	10	4	11	3	12
23	5	10	5	11	4	12	4	12
24	6	11	5	11	4	12	4	13
25	6	11	5	12	4	13	4	13
26	6	11	6	12	5	13	4	13
27	7	12	6	12	5	13	5	14
28	7	12	6	13	5	14	5	14
29	7	13	6	13	6	14	5	15
30	8	13	7	14	6	15	6	15

A5. Critical Values for Symmetry Test for $\alpha = 0, 10$

n	m_1	m_2	n	m_1	m_2
5	1	4	18	5	13
6	1	5	19	5	14
7	1	6	20	6	14
8	2	6	21	6	15
9	2	7	22	7	15
10	2	8	23	7	16
11	3	8	24	7	17
12	3	9	25	8	17
13	3	10	26	8	18
14	4	10	27	9	18
15	4	11	28	9	19
16	4	12	29	9	20
17	4	13	30	10	20

A6. Critical Values for Maximum Series Length Test for $\alpha = 0, 05$

Number of residuals (n)	Maximum series length k_t
10	5
14	6
22	7
34	8
54	9
86	10
140	11
230	12

A7. Critical Values for Number of Series Test for $\alpha = 0,05$

	n_1																		
n_2	2	3	4	5	6	7	8	9	10	11	12	13	14	15	16	17	18	19	20
4			2																
5		2	2	3															
6		2	3	3	3														
7		2	3	3	4	4													
8	2	2	3	3	4	4	5												
9	2	2	3	4	4	5	5	6											
10	2	3	3	4	5	5	6	6	6										
11	2	3	3	4	5	5	6	6	7	7									
12	2	3	4	4	5	6	6	7	7	8	8								
13	2	3	4	4	5	6	5	7	8	8	9	9							
14	2	3	4	5	5	6	7	7	8	8	9	9	10						
15	2	3	4	5	6	6	7	8	8	9	9	10	10	11					
16	2	3	4	5	6	6	7	8	8	9	10	10	11	11	11				
17	2	3	4	5	6	7	7	8	9	9	10	10	11	11	12	12			
18	2	3	4	5	6	7	8	8	9	10	10	11	11	12	12	13	13		
19	2	3	4	5	6	7	8	8	9	10	10	11	12	12	13	13	14	14	
20	2	3	4	5	6	7	8	9	9	10	11	11	12	12	13	13	14	14	15

Critical values of k_2

n_2 \ n_1	2	3	4	5	6	7	8	9	10	11	12	13	14	15	16	17	18	19	20
2	4																		
3	5	6																	
4	5	6	7																
5	5	7	8	8															
6	5	7	8	9	10														
7	5	7	8	9	10	11													
8	5	7	8	10	11	12	12												
9	5	7	9	10	11	12	13	13											
10	5	7	9	10	11	12	13	14	15										
11	5	7	9	11	12	13	14	14	15	16									
12	5	7	9	11	12	13	14	15	16	16	17								
13	5	7	9	11	12	13	14	15	16	17	17	18							
14	5	7	9	11	12	13	15	16	16	17	18	19	19						
15	5	7	9	11	13	14	15	16	17	18	18	19	20	20					
16	5	7	9	11	13	14	15	16	17	18	19	20	20	21	22				
17	5	7	9	11	13	14	15	16	17	18	19	20	21	21	22	23			
18	5	7	9	11	13	14	15	17	18	19	19	20	21	22	23	23	24		
19	5	7	9	11	13	14	15	17	18	19	20	21	22	22	23	24	24	25	
20	5	7	9	11	13	14	16	17	18	19	20	21	22	23	24	24	25	26	26

Index

© Springer International Publishing AG 2018
J. Welc, P.J.R. Esquerdo, *Applied Regression Analysis for Business*,
https://doi.org/10.1007/978-3-319-71156-0

Zeitfracht Medien GmbH
Ferdinand-Jühlke-Straße 7
99095 Erfurt, Deutschland
produktsicherheit@kolibri360.de